Technology Policy Meets the Public

PESTO Papers 2

edited by Andrew Jamison

Technology Policy Meets the Public
PESTO Papers 2

©   Andrew Jamison
ISBN   87-7307-611-2

Published by:
Department of Development and Planning
Aalborg University
Fibigerstræde 13
DK-9220 Aalborg Øst, Denmark

Distribution:
Aalborg University Press
Badehusvej 16
DK-9000 Aalborg, Denmark

Phone: +45 9813 0915
Telefax: +45 9813 4915

All rights reserved. No part of this book may be reproduced, stored in a retrieval system, or transmitted, in any form or by any means, without a reproduction license or the permission of the publisher, except for reviews and short excerpts in scholarly publications.

Layout by Bente Vestergaard
Printed in Denmark 1998 by Thy Bogtryk & Offset A/S, Thyholm

Contents

	Page
Foreword	5
1. Sustainable Development and the Problem of Public Participation *Andrew Jamison and Brian Wynne*	7
2. Making Participation Happen: The Importance of Policy Entrepreneurs *Marco Giuliani, Leonardas Rinkevicius and Arni Sverrisson*	19
3. Representing the Public: New Roles for Environmental Organizations *Kees Dekker, Mario Diani, Andrew Jamison, and Lise Kvande*	49
4. Participation by Mandate: Reflections on Local Agenda 21 *Jose Andringa, Marco Giuliani, Patrick van Zwanenberg, and Magnus Ring*	81
5. Roads to Sustainable Transportation? On Public Engagement in Infrastructure Projects *Patrick van Zwanenberg, Robbin te Velde, and Per Østby*	109
6. Environmental S&T Policy in One Country: The Public-Policy Interface in Italy *Marco Giuliani*	147
7. Innovation Concepts and Cleaner Technologies: Experiences from Three Danish Action Plans *Arne Remmen*	173
8. Public Engagement and UK Agricultural Biotechnology Policy *Patrick van Zwanenberg*	189
9. Constructive Technology Assessment Comes of Age *Johan Schot*	207
Contributors	233

Foreword

These papers have been written as part of the research project, Public Participation and Environmental Science and Technology Policy Options (PESTO), supported by the European Union's program in targeted socio-economic research. They are based on research conducted during the first "work package" of our project, which in most of the participating countries was carried out from September 1996 to December 1997. In the same period, there has been a slightly different Nordic project, which has been conducted in conjunction with the EU project, and we acknowledge the support of the Nordic Environmental Research Program in making possible the publication of our results in this form. These papers are to be seen as a joint product of the two overlapping project groups (research teams in Sweden, Norway, and Lithuania are partners in both projects; Italy, the Netherlands and the United Kingdom are only participants in the EU project, while Iceland and Denmark were only partners in the Nordic project).

It should be stressed that the papers are exploratory in nature, since the time available has been quite limited and each author has approached the material from somewhat different perspectives. There has been a certain common terminology and, most importantly, a common set of questions that have been asked of the material, based on discussions at PESTO meetings in Trondheim (November 1996) and Aalborg (May 1997). At those meetings, we agreed that it would be valuable to seek to divide the multidimensional "public" into more specific categories, and to discuss particular issues, such as the different types of entrepreneurship that have emerged to mediate between the public and the policy makers, the changing nature of activities carried out by environmental organizations, the varieties of experience with local Agenda 21, and the role of the public in sustainable transportation programs. The chapters that make up part one of this book were then written by a team of authors, drawing on examples from at least two of the different partner countries.

In part two, we present single-authored papers which were originally intended to be included in collaborative chapters. They are presented here, since their quality and informative value warrant wider dissemination. One presents the experience of public-policy interaction in Italy, a second the shifts in the Danish policy discourse on cleaner production, a third extends the PESTO perspective into the area of biotechnology and the final chapter discusses constructive technology assessment as a way to include the public more directly in technological development.

The production of this volume has been an experiment in both the form and content of research collaboration, and I would like to thank the authors for their patience and cooperation in a process of collective writing.

This anthology is the second in a series, and it might be of interest for readers to consult the first set of PESTO papers, which were published by Aalborg University Press in 1997 under the title *Public Participation and Sustainable Development: Comparing European Experiences*. Future publications are planned, and interested readers are invited to contact me if they would like additional information about PESTO. As editor and project coordinator, I thank my colleagues in the project for their contributions, and for their ability to translate my project vision into real results. I also wish to thank Bente Vestergaard for her able assistance in the further translation of those research results into readable form. For the more direct translation of material into English, I thank Richard Rogers and Emily Jamison Gromark.

Aalborg in May 1998
Andrew Jamison

Chapter One

Sustainable Development and the Problem of Public Participation

by Andrew Jamison and Brian Wynne

1. The Environment as a Bonding Narrative

In just about every industrialized country, it was at some point in the 1980s that environmental concern ceased to be a living source of collective identity for a relatively small number of movement activists and became instead a societal discourse: a bonding narrative. The apocalyptic tones, the "bad news" that had characterized so much of the environmental debate up till that time was gradually transformed into the encouraging, good news rhetoric of sustainable development, which, since it was propounded in 1987 in the Brundtland report, *Our Common Future*, has provided inspiration for very different kinds of social actors. What had previously been a wide ranging critique of industrial society and its waste and artificiality has been more or less replaced by a much more delimited set of symbols, ideas, slogans and practices that, in the 1990s, have been working their way into the worlds of science and technology, of business and government. The environmental "movement", which had earlier been seen by those in power as a threat to the further expansion of the corporate state has come instead to be seen, by many actors in both business and government, as an important contributor to economic recovery and rejuvenation, as well as a participant in developing new forms of scientific and technological knowledge (cf. Eder 1996, Jamison 1996).

From the paradigmatic notions of sustainable development and ecological modernization to the pragmatic techniques of cleaner production and pollution prevention to the new marketing strategies of green consumption and environmental labeling, the discourse of environmentalism has been reinvented over the past ten years as a cluster of green competences, as new forms of environmental expertise. The critical rejection of the wonders of modern science and technology that the environmental movement articulated in the 1970s has come to be deconstructed and reconstituted, from the mid 1980s onward, as central components in constructive programs of science, technology and economics.

This discursive shift is intimately connected to changes in the character of the international political economy. By the mid 1980s, socio-economic life had become increasingly "globalized", with research and invention often carried out in one part of the world, technological innovation and development in another, and manufacture in still others. Individual firms had increasingly become nodes in transnational corporate networks, and socio-economic relations in general had more and more come to be governed by international institutions and rules of behavior. Both in terms of production and consumption, the fundamental structures of organization and decision-making had moved to a transnational space, making it increasingly difficult for nation states and governments to impose, or even articulate, their own independent policy agendas.

These geographical transformations have been accentuated – and, according to some observers, in large measure caused – by developments in telecommunications and information technology. In the 1980s, it became possible, and, in a few short years, common practice, to carry out economic transactions and conduct industrial activities on a global basis, and to shift operations from country to country depending on changes in market and financial conditions. In Europe, these developments have fed into the efforts to integrate policy making and to develop new kinds of institutions at a European "level". Increasingly, economic activity is conducted across national boundaries, and the key policy functions have been taken over by European regulations, commissions, authorities, and agencies.

For environmentalism, and environmental science and technology policy, these transformations have meant a shift in substantive focus – from the local and national to the transnational and global, when it comes to the issues to be dealt with – as well as a shift in location – from national bodies to intergovernmental and international organs, when it comes to policy making and implementation. In actual research practice, the new information and communication technologies have meant a great deal, in terms of the kinds of observations that can be simulated, the kinds of models that can be constructed, and the kinds of calculations that can be made. The "social construction" of scientific facts has been shifted from a more or less direct interaction with the environment and its component parts, to an ever more abstract and aggregate virtual environment of atmospheric, hydrological and geological processes that cannot be directly observed or, for that matter, studied.

It is increasingly apparent that the new environmental problems require for their "solution" something more than traditional, disciplinary science and technology. They rather seem to call for what has been termed a new "mode of knowledge production" that combines different disciplinary perspectives in a problem-oriented and context-dependent transdisciplinarity (Gibbons et al 1994). In particular, there is need for an intermediary expertise between the social and the technical, an expertise in environmental social science, or, as it is often called, the human dimensions of global change. This expertise involves a knowledge of particular methods of accounting, assessment, scenario building, forecasting, foresighting, and prediction for dealing with the extremely abstract and uncertain new range of environmental problems. But it is also, at various levels and in various ways, an expertise in societal adjustment, environmental management, life-cycle analysis, risk assessment. It is what Ulrich Beck terms reflexive knowledge, a form of knowledge that Beck sees as characteristic for the "risk society" in which we in Europe now find ourselves (Beck 1992; cf. Beck 1995).

The broad change of focus involved in the discursive move from "environmental protection" to "sustainable development" has complicated our appreciation of the attendant challenges, by explicitly proposing that the issues are much more than those of maintaining the functional viability of the natural environment. This has given widespread recognition to the insight that the natural and the human are inextricably intertwined and mutually defining. But the dominant "modernist" discourses in public policy, as well as in social theory and planning, continue to find it difficult, if not impossible, to digest and take seriously this fundamental insight. Thus the recognition-in-principle of the human challenges posed by sustainability still finds those human dimensions inadequately – indeed often counterproductively – expressed in the doctrines and programs of public policy.

It has been increasingly acknowledged that contemporary environmental challenges require for their resolution more than mere technical "fixes" to achieve optimal eco-efficiency in all areas of social and economic life. Even though finding appropriate technical solutions is important and challenging in itself, it has been fairly widely accepted that even the most optimistic of scenarios of such putative eco-efficiency improvements – which will have to be not just innovated but implemented, too – will be overtaken in terms of environmental impact, due to inexorably increasing consumption. In other words the background problem is our apparently still unrestrained and expanding demand for goods and services, with the attendant strains on natural systems. If eco-efficiency – and a greener, cleaner production and consumption on all fronts – were a realistic prospect, it might buy us time, but it does not obscure the fundamental conclusion that unlimited material human demands and aspirations, as reflected in our basically exploitative relationships with nature and each other, are the main problem. These are problems of our human subjectivities, and of human culture, which therefore cannot be left to be reproduced and reinforced by default.

2. The Resistance of the Established

The quest for sustainable development has come to provide a new collective heuristic, a new historical project for late modern societies. The interpretive flexibility and lack of precision in such concepts as sustainable development, cleaner production, environmental management and green consumption offer opportunities for diverse actors to be held together in a common discursive arena, even though the particular policy options continue to be highly contested. At a time when the role of the state is being diminished in many areas, thus calling into question traditional notions of democratic process and representation, the quest for sustainability has led to an array of new forms of knowledge production and to a search for new modes of public participation. Indeed, a broader public involvement has come to be seen as fundamental to the effectiveness of many of the new policy proposals and measures.

Despite a certain rhetorical association between environmental sustainability and democratic renewal, however, the dominant forms of public policy discourse continue to serve as obstacles for broader participation. On the one hand, there has been a transfer of responsibility in many areas of environmental science and technology policy from the public to the private sectors, and a decrease in direct state involvement in research and innovation. Privatization has tended to limit public access to decision-making and to the setting of policy agendas. On the other hand, the typical form of policy making privileges technical expertise; in problem formulation, as well as in policy implementation, an instrumental and objectivist mind set delimits human agency, and tends to reduce social and political issues to matters of technical measurement and expert evaluation. Even non-governmental organizations are often affected by this scientistic cultural bias, seeking niches in the policy arena for instrumentalized and professionalized versions of environmental competence.

The instrumental forms of knowledge which have been virtually a defining feature of policy and expert advice embed and reproduce existing conceptualizations of the human subject and our instrumental relationship with nature. Thus whatever use they have been thus far, they need complementing and indeed often challenging as to the proper and desirable scope of the instrumental ethic embedded within them. This instrumentalism does not pertain only to natural-scientific or technological forms of knowledge, which have been defined by the epistemic principle of instrumental prediction and control at least since the 17th century scientific revolution. It also pertains to dominant social sciences, especially economics, which play a defining role on the public policy domain. These instrumental and behaviorist social sciences such as most of psychology, geography, economics, demography and many others are epistemically correspondent with the natural sciences in these respects, humans being analytically constructed as objects which follow deterministic laws of behavior.

Even within scientific environmental or risk public policy discourses that are apparently only about nature, and nothing to do with the human, implicit models or visions of the human do thread such supposedly purely propositional "natural" knowledges. This hermeneutic dimension is always latently present, but especially when such natural knowledge is explicitly used as policy authority and the boundaries between scientific knowledge and policy as ambiguous as they typically are.

In his 1986 book, Ulrich Beck described processes of individualization going on in the mainstream institutions of modern society, fragmenting them from within and destroying the individual's identification with the institution – work, family, educational base, political party, etc. Those institutions correspondingly no longer offer protection, loyalty or commitment to which the individual might reciprocate. Against this backdrop the further factor of environmental risks intensifies these dynamics, according to Beck, and gives them their fundamentally new and distinctive form. These risks, he argues, are generated by modern science and its institutions, yet are no longer contained and controlled by them. Unlike in previous eras, even the rich and powerful cannot escape them. They are now pervasively global, uninsurably large and catastrophic in potential, and irreversible. Thus modern science, the epitome of modernity, has created a monstrous and comprehensive risk situation, yet cannot manage it. Even worse, according to Beck's thesis, scientific institutions cannot summon the integrity or maturity to acknowledge and take responsibility for this dire and historically new predicament.

Faced with this reality with its central breakdown of the scientifically-inspired maintenance of civil security, as Beck describes it, citizens at large have withdrawn identification, trust and legitimacy from modern scientific and expert-led institutions. Modernity as such has taken a reflexive turn, as ordinary people question the basis of political and technical authority which is constantly embarrassed by unanticipated, and often even denied, ecological and human health consequences of its own previous "scientific" actions and technologies. People instead identify with and commit to new informal, extra-institutional forms of political activity often focused around issues previously defined as unpolitical, like lifestyle, health, and cultural practices; hence the growth of new subpolitical spheres and movements and cultural interest-groups of myriad kinds actively hostile to conventional institutional politics and policy.

Anthony Giddens' version of this reflexive process of sociocultural change in what he calls "high modernity" contains some key differences but also similarities with Beck's account (Giddens 1991). He emphasizes more the rise, in every walk of life, of expert disagreement and uncertainty (a version of expert institutions' inability to control risks), and the lay public's unprecedented encounter with a radical existential need therefore to make life-identity choices (including, crucially, "which experts shall I trust?") previously taken care of by monolithic (and according to Giddens, trusted) expert institutions. Giddens shares with Beck a concern with globalization, and with the severity and irreversibility of risks, but stresses not the role of ignorance (unanticipated environmental effects) in generating public mistrust, but the self-reflexive knowledge of the modern scientific temper itself as it has diffused more widely in modern society.

3. New Forms of Representation

For all their originality and persuasiveness in many respects, and despite some fundamental differences, for example as to the agent of the reflexive force at work (ignorance or knowledge? nature or human?), both these theories share an embedded construction, or representation, of the public as responding through a calculative, instrumental and individual subjectivity. The cultural, human responses to the inadequate forms of representation of the human given in rational policy discourses are inherently difficult to demonstrate, and are themselves a process recognized by our considered but fallible interpretive intervention, not by pretended observation alone. They advance a model of the human which is less reduced to a calculative instrumental individual and more of a human subject animated by non-instrumental feeling as much as instrumental calculation, and relational rather than isolated.

Natural discourses in public policy can never be purified of human correspondences which take on normative dimensions, but they should – and can – be rendered more transparent and publicly accountable. It seems that problematic representations of the human are being exercised but at the same time buried from open view in modern expert-led policy discourses about environment and risk. Moreover the kinds of tacit assumption, projection or representation of the human are not simply hypothetical models which are cast upon the waters of public debate and response to be tested and if inadequate, revised or replaced. They are typically not even recognized as existing and influencing public self-understandings, because scientific discourses are vehemently defended as exempt from any such human dimensions. Thus by default of such acknowedgement they become not merely representational errors, but an ontological program which in effect potentially imposes them as normatively authoritative definitions of the human dimensions of such issues and indeed more widely. Thus the imagined "human dimensions" may gradually – not deterministically but subtly – materialize as human subjects adapt to their intensified diffusion in public culture.

If such tacit human representations are inadequate as emotionally and intellectually recognizable discourses, they are not open to correction by purely intellectual means because they do not identify themselves in this way – they are tacit, and maybe even not conscious or deliberate, just reflecting available cultural resources in the prevailing policy and disciplinary discourses. Thus public reactions to the possible inadequacy of such human representations are likely to be indirect, and culturally-practically based rather than intellectual. They are most likely to engender disaffection, alienation, lack of moral identification, mistrust and practical self-differentiation from them without this necessarily being explicitly rationalized, deliberated, and "chosen" through conscious decision. The result is a culturally rooted, humanly engendered response to what may often (though not always) be a diffuse sense of profound alienation from the implicit representations of the human in dominant discourses which we are expected to respect and grant authority to.

Such an interpretation of a basically cultural process of public recoil and alienation from expert-led rational policy making and surrounding debate (for example in the ways the public is represented in discourse of public understanding of science, or in surveys of attitudes to environmental risks and science) is entirely consistent with the widespread research finding and common experience of public mistrust of and disaffection from modern forms of policy discourse on risks, environment and related issues (cf. Macnaghten and Urry 1998). This is not to argue that the posited poverty and concealment of such tacit human representations in official discourses is the only reason for this observed syndrome of public mistrust, but that it is an important factor. The cultural problem of unreflective embodiment and dissemination of such inadequate human representations is bound to make the problem of public alienation even worse, as research has shown that people are realistic enough to recognize their own unavoidable dependency upon those same expert institutions which are both so badly misrepresenting them, and also avoiding recognition of those open and debatable human dimensions. Open recognition of this ultimately conditional nature of their scientific knowledge by such expert institutions would be a prior condition of their public authority and legitimacy. Yet they still appear to exercise an anachronistic contrary cultural instinct that their public authority depends upon the concealment of any such indeterminacies underlying their explicit natural expert discourses.

The depth and pervasiveness of this problematic and self-defeating modern cultural syndrome is not adequately taken into account in the social theories of modern cultural change of Beck and Giddens. Though these have provided a wealth of insight into late-modern conditions, they have failed in respect of their unreflexive adoption of a decisionistic, rational-choice and instrumental model of the human. Yet we must be aware that the forces and feelings engendering cultural, as distinct from instrumental and chosen, responses of public alienation and disaffection from modern expert policy systems or actors, are not going to be directly expressed or even expressible, hence they are never going to be amenable to direct demonstration in fieldwork situations, however intimate. It is also worth noting that such mistrust and alienation is often ambivalent, tinged with realism about dependency, and shows a countervailing conditional readiness to "trust" expert institutions. This may also reflect the essentially open and indeterminate interpretive work which the public also has to do to apprehend the tacit human representations which are problematic within expert discourses.

To summarize, the three basic aspects of most of the modern discourses of expert policy institutions which are problematic are their:

- determinism, which both symbolically and materially imprisons the human subject in a web of projected natural determination, predictability and controllability;

- their empirically and morally inadequate substantive accounts of the human as instrument, individual pursuers of self-interest;

- and the fact that, whether or not they are adequate substantively, they are hidden and unacknowledged, thus unaccountable and profoundly antidemocratic given that they potentially shape material human dispositions of a political and ethical kind.

4. Exploring the Public-Policy Interface

As we shall see in the following chapters, the actual modes of "public participation" that have emerged in recent years – from local Agenda 21 activities to hearings and consensus conferences and innovative approaches to technology and environmental impact assessment – are highly fragile and, in many countries, appear to be disconnected from the real sources of power and decision-making. While serving to construct new forms of consciousness and raising awareness about connections between different environmental problems, as well as between environmental and broader social welfare issues, the participatory experiments are often temporary. At the same time, seeds for new forms of representation are being planted, but their implications are highly ambiguous. Embryos of new public interest are perhaps developing in green lifestyles, as well as new forms of local-based "subpolitics" which have not yet had any meaningful connections with formalized, established forms of politics. What "publics" are actually being represented is, however, still largely indeterminate (cf. Lash et al 1996; Szerzynski 1997).

Two distinct strategies appear to be emerging in relation to sustainable development, with characteristic patterns of public participation. On the one hand, in the various technocratic projects of so-called ecological modernization, participation is primarily conceived in a top-down way, with the public given the role of the environmentally-conscious consumer, offered opportunities for ecological employment and the participation of the pocketbook. On the other hand, as opposed to this are the bottom-up approaches emanating from locally-based initiatives, where forms of participation remain open-ended and highly diverse. The pursuit of environmental sustainability provides a catalyst for experimentation with new forms of sociality and association. And, particularly when combined with cultural events and aesthetic expression, modes of participation can take on the character of "exemplary action", performing and disseminating sustainability through telling example or model behavior. (See Eyerman and Jamison 1998)

In the chapters that follow, we approach these issues by attempting to problematize the roles of the public in relation to sustainable development. Public participation, or civic engagement, are terms that are often invoked by policy makers, but seldom reflected upon in a particularly serious or systematic fashion. The public is supposed to be involved in decision making: such is the rhetoric of democracy. It is on behalf of the public, however amorphous or abstract that public might actually be in reality, that decisions are made, agendas and new doctrines are formulated and programs and projects are implemented. But how does the public manifest its multifaceted interests in the making of policy decisions?

In academic discourse, the public is a vague, even contradictory, entity, continually reinventing and redefining itself, organizing itself in new constellations – movements, interest groups, political parties, non-governmental organizations – and repeatedly imagining new "roles" for its own various component parts to play. What is to be meant by participation? What influence, if any, is the public allowed to have over the deliberations of governments and parliaments? What forms of involvement are culturally acceptable and which are not? These are extremely difficult, but at the same time extremely important, questions to try to answer. At a time when policy decisions in many areas of public life are becoming ever more globalized and the distance between the public and the policy makers is generally increasing, it is a central democratic task to address the problem of public participation.

In relation to science and technology policy, the public has an especially difficult time, for this is an area usually reserved for experts, and the opportunities for the public to make its multifarious voices heard are even more limited than in other areas of policy making. In order to analyze these processes, we have drawn on different kinds of theories and conceptual frameworks and explored different examples of public participation, or civic engagement. We have had what might be termed a weak comparative ambition, examining similar processes with somewhat different methods, levels of abstraction and styles of presentation. Due to the nature of the PESTO exercise, which, as part of the EU program on targeted socio-economic research, does not provide time or resources for in depth empirical research, the papers cannot be definitive. Rather, they are meant more to serve as illustrative examples of research strategies and approaches – of what could be investigated and how – rather than as finished analyses. It is well known that the manner of public involvement in policy making differs from country to country, and is determined by a range of factors that are difficult to disentangle, and our effort here can obviously offer no final assessment of the problem of public participation. But we feel that the papers do provide a number of new insights, and we present them in this form so that they might be able to make some modest contribution both to the theory and analysis, but also to the practice and improvement of public participation.

References

Beck, U. (1992, 1986) *Risk Society. Towards a New Modernity.* Sage.
Beck, U. (1995) *Ecological Politics in an Age of Risk.* Polity.
Eder, K. (1996) 'The Institutionalisation of Environmentalism: Ecological Discourse and the Second Transformation of the Public Sphere', in Lash et al.
Eyerman, R. and A. Jamison (1998) *Music and Social Movements*, Cambridge University Press.
Gibbons, M., et al. (1994) *The New Production of Knowledge.* Sage.
Giddens, A. (1991) *Modernity and Self-identity: self and society in the late modern age.* Polity.
Jamison, A. (1996) 'The Shaping of the Global Environmental Agenda: The Role of Non-Governmental Organisations', in Lash et al.
Lash, S., B. Szerszynski and B. Wynne, eds, (1996) *Risk, Environment and Modernity. Towards a New Ecology.* Sage.
Macnaghten, P. and J. Urry (1998) *Contested Natures.* Sage.
Szerzynski, B. (1997) 'The Varieties of Ecological Piety' in *World-Views* 1: 37-55.

Chapter Two

Making Participation Happen: The Importance of Policy Entrepreneurs

By Marco Giuliani, Leonardas Rinkevicius and Arni Sverrisson

1. Introductory Observations

The environmental agenda is no longer limited to regional or national issues: globalization and sustainability have gradually become the keywords of the green policies of the 1990s. This new construction of the problems calls for an internationalization of the policy instruments, whereas an anticipatory pro-active approach is the unavoidable corollary of the need for comprehensive strategies and planning.

The strive towards sustainability also demands new planning capabilities in a period in which the regulatory role of the state is more and more disputed. It calls for a collective engagement around an issue which is beset with all the paradoxes of public goods and inter-generation cost transfer. Simultaneously, governance problems are shifted to a transnational level, not least within Western Europe, decreasing the possibilities of national actors to create and control a political consensus, and investing relatively weak supra-national organizations with new responsibilities (Dente 1994). The result is fluid jurisdictions and uncertain futures, in contrast to Max Weber's ideal-typical bureaucracy. Among the issues thus decided increasingly at the political rather than the administrative level is the content and practicalities of sustainability.

In this context, policy entrepreneurs with environmental agendas present new opportunities as well as problems. They tend to operate across the traditional boundaries between public and private, politics and administration, state and market, consensus and profit, expertise and representation. In their activities, they both bend existing administrative and representative mechanisms to their own ends, while simultaneously they question their effectiveness in meeting the challenge of ecological modernization. How this trend is manifested in different countries is the main subject of this chapter.

First, we offer some general observations on policy entrepreneurship. We then develop a typology of the practices characterizing policy entrepreneurship. A discussion of policy entrepreneurship in Italy, Lithuania and Sweden follows and in our concluding section, we outline our research findings and further research issues.

In what follows we will explicate the defining attributes of this type of actor in order to answer the question: what is a policy entrepreneur? We will also analyze their role in the recent evolution of environmental policy making, and particularly how environmental policies have started to "trespass" on the domain of science and technology policy as traditionally understood. What do policy entrepreneurs do in this particular context?

We will then specify the endogenous and exogenous conditions of policy entrepreneurship, and discuss what kind of resources they activate and which type of contexts facilitates this mode of operation. As we go along, we will also illustrate the role of policy entrepreneurship in the problematic symbiosis currently developing between environmental policy and science and technology policy as the agenda of sustainable development, such as it is, is being implemented.

Science and technology policy can be conceptualized as an arena of interaction between four policy domains, or constituencies – economic, bureaucratic, academic, and civic. These domains are characterized by different ideals of science and technology policy, that is, by different attitudes to the general social functions of science and technology. Each domain also tends to favor particular kinds of policy measures, as well as different types of programs and projects (Elzinga and Jamison 1995). This general perspective is summarized in the following table.

Policy domain	Bureaucratic	Economic	Academic	Civic
Doctrine	order	growth	enlightenment	democracy
Steering mechanism	planning	commercial	peer review	assessment
Ethos	formalistic	entrepreneurial	scientific	participatory

The following should be read as a part of an attempt to analyze the intersections of the science and technology policy cultures as conceptualized above. The different cultures are manifested in government agencies concerned with environmental issues, in companies working in "green" niches or struggling to present their production as "green", in academic institutions involved in applied, strategic and basic research relevant to environmental issues, and in environmentalist movements and organizations. Each of those can be platforms for policy entrepreneurship, that is, generate new approaches which are then diffused and shape the environmental science and technology policy agenda in their respective countries. If this is to take place, however, alliances must be created and credibility, then legitimacy, and lastly an interpretative primacy gained for the proposed paradigm.

In this context it is important to note that the social space in which the idea of sustainable technology has been born and spread is not, we have maintained, constituted by a single network in which contacts and connections are evenly spread but rather a field of heterogeneous clusters which do however interact with each other in various ways. This, we should note, implies a preconception of environmental science and technology policy as an exceedingly fragmented field of interaction. On the one hand, technological ideas which lead to less waste or cleaner production methods sprout within academic institutions or in company R&D departments. Sometimes, but not always, they are taken up by environmental organizations or government agencies and pursued as part of sustainability policies. Political actors, however, including parties, top-level bureaucrats and movement organizations, tend to start from the other end, that is pose environmentally oriented goals the implementation of which may be specified in some detail or largely left to the "experts". In either case, specific efforts are usually needed for promoting particular ways of conceptualizing the link between sustainability and a particular technological solution and in all the countries discussed here there is no single channel for such efforts. Rather, networks are built, on an ad hoc basis, but usually not from scratch: rather, existing networks, positions and other resources are *mobilized* and *reconfigured* in order to acquire credibility, legitimacy and eventually, interpretative primacy. The opportunities for doing this usually arise on the occasion of particular events, campaigns, law proposals when persons who in the ordinary course of events do not have regular contacts are brought together to pursue specific goals, or contest them.

Hence, policy entrepreneurs can be seen as a particular type of entrepreneur, who establishes connections where none existed before among actors with different backgrounds, or develops existing weak relations, changing their content, with the express aim of influencing policy. In our case this means building bridges between environmentally oriented actors and those traditionally concerned with science and technology policy, with a view to developing a mutual understanding, in order to work towards a shared goal. Such strategically placed entrepreneurs who create, amplify and maintain links among groups located within different policy domains, creating what we can call sustainable technology networks, are needed exactly because the "normal" networking based on background and trust etc. can be expected to be insufficient in this particular case. The networking based on common attributes, shared meanings etc. would tend to occur mainly within policy domains, if left to itself. Put more simply, activists, academics, corporate leaders and politicians tend to talk to or even at each other and not with each other.

However, if successful, such networks can be expected to coagulate, for example in the form of projects which are later turned into permanent organizations, which may include the whole original network or be the result of a focusing process where redundant ties are cut off (cf. Burt 1992). The resulting unit is likely to be firmly established within one policy culture or another, but retaining ties to some of the other cultures as a normal part of its operations.

We can identify two different types of action performed by policy entrepreneurs. On the one hand, there is cognitive/discursive action, which is aimed at promoting new perspectives, identifying original solutions, shifting the common perception of an issue and forging the debate around it. However, as other entrepreneurs, policy entrepreneurs must operate within the constraints implied by the state of the art: in other words, they mainly promote and diffuse "ideas in good currency" (Milward 1980: 263), which are potentially "powerful for the formation of public policy" (Schon 1971: 123). Within this general constraint, they can address the choice of policy instruments, some technical solution still undervalued, the organizational setting of policy-making, or the philosophy of public intervention. The changes proposed by policy entrepreneurs may thus lie at the core of the policy-making process, or be peripheral and instrumental, but they all tend to induce changes in the means-end relationship which underlies the normative frameworks of environmental and science and technology policy making (Svenning 1996).

The other type of action performed by policy entrepreneurs aims at building coalitions between policy actors. Particularly interesting for us are policy entrepreneurs who link actors who traditionally belong to different camps: industrialists and environmentalists, left and right wing parties, Labor and Greens, conservatives and reformists, common citizens and ecological activists, etc. This type of straddling (and struggling) alliances appears time and again in efforts towards promoting the diffusion of environmentally oriented technological practices, as we shall see below. Further, to overcome the intrinsic limits of collective action the presence of a catalyst which unites scattered interests is vital. Analyzing regulatory efforts, it has been recognized that policies which produce the peculiar combination of concentrated benefits and dispersed interests need some kind of political entrepreneurs to be approved:

> their defining characteristic ... is their willingness to invest their resources – time, energy, reputation, and sometimes money – in the hope of a future return. That return might come to them in the form of policies of which they approve, satisfaction from participation, or even personal aggrandizement in the form of job security or career promotion (Kingdon 1984:129).

Their linking activity may be pursued both explicitly, trying to attract the support of diverse interests in order to establish a firm network, and "unconsciously", simply providing the means for mutual recognition and sharing of information by clusters of actors otherwise weakly connected. Thus, the introduction of environmental standards may not need this type of coordinating activity, but their implementation in a variety of contexts does. The adoption of complex programs like, for example, the implementation of the Agenda 21 goals at a national or local level is also probably favored by broader webs of relationships.

Because policy-making is both a strategic/organizing and cognitive/discursive activity in the sense outlined above (cf. Majone 1989), the outcomes are strongly dependent upon the capacity of leading actors for actually tying together an adequate number of formal supporters – politicians, bureaucrats, interest-groups, stake-holders, etc. However, they are even more sensitive to the development of common interpretations among different publics. These interpretations or story-lines (Hajer 1995) must, in order to be effective, draw together experts and policy-makers, activists and lay-people. Policy entrepreneurs consequently act at both these levels by constructing alliances *and* molding policy discourses. Their contribution appears particularly relevant whenever they succeed to dismantle petrified (op)positions and open the road to new alliances through the reinterpretation of long standing dilemmas. Hence, policy entrepreneurship becomes particularly important in providing paradigms for policy making which reconcile the need for regulation and the profit motive, state control and market mechanisms. This is typical of the new types of environmental policies now emerging, which address the ways in which companies manage production and distribution processes, and the development of technologies used in such processes. Demonstrating that zero-sum games can be positive-sum games, that "green and clean" can be cheaper as well, thus neutralizing old cleavages, is among the main functions of policy entrepreneurship in the environmental field today.

Let us preliminarily summarize the characteristics of policy entrepreneurs. First, they are *leaders* who are recognized as such by other actors. Second, they are *innovators*: they do not administer routine processes but foster reforms and advocate change. Third, they are *catalysts,* who mobilize latent and manifest networks in order to promote environmentally oriented paradigms within the field of science and technology policy. Lastly, they are participants in public debates, interpreting the issues as well as the discursive consensus within particular coalitions to their own ends.

The first step for policy entrepreneurs is to acquire credible positions, to be taken seriously, as it were. Lay persons complaining about emissions, posing relationships between these and particular diseases, are for example routinely treated as non-credible (Irwin 1995), and the solution to this problem is usually the mobilization of scientific expertise, hence creating a cleavage within the academic domain itself. Endorsement by a variety of political groups also creates room for maneuver.

The next step is to contest the legitimacy of established positions and attempt to acquire legitimacy for one's own. Legitimacy refers mainly to the legitimacy of governmental authority, which, however, is not monolithic: hence legitimacy can be acquired through the action of municipal authorities, through the action of government agencies, through policy statements by powerful parties etc. However, the legitimacy of democratic politics is not limited to actors within the state. Rather, it is a question of legitimate participation in the political process. Thus leaders of recognized movement organizations have a legitimate position, as do spokesmen of academic organizations, trade unions, and so on. Within and around any administration, resources of one or more actors can in other words be drawn on against those of others, a consequence of the fluid jurisdictions noted above and of the expansion and heterogeneous character of the modern state, and non-governmental actors will therefore in early stages reach out to bureaucratic and political actors and attempt to include them in their alliances, and by accepting the "rules of the game", obtain and maintain legitimate positions as spokespersons representing particular interests.

Conquering interpretative primacy is the third step, and requires not only authoritative action but also consensus among the public and major media about the relevance and rightfulness of the position adopted. In Bourdieu's terms new positions need to be recognized as interpretations of the reigning doxa or in other words, be transformed from a heterodox to an orthodox position (Bourdieu 1980).

Although a policy entrepreneur will probably derive her credibility mainly from one of the policy domains referred to above – namely the bureaucratic, the economic, the academic and the civic domains – her legitimate authority to speak on environmental issues and their implications for science and technology policy will have to be recognized by a wider public. No single resource is likely to be sufficient to establish a legitimate position but below we discuss a number of resources which can be drawn upon for this purpose.

First of all, though holding some *formal position* of responsibility is neither a defining nor a necessary condition for entrepreneurship, still it may reveal itself as a powerful resource which can be activated and exploited by policy entrepreneurs. Top executives, like the Minister for the Environment, or the head of a regional department for environmental affairs, are certainly bound to the bureaucratic organization of their office. Nevertheless it is possible to observe the innovative turn impressed on their office by top executives highly committed to environmental goals.

The same may happen in the economic domain. New generations of business-people and engineers exhibit a growing concern with the environmental effects of their industries, manifested in a concrete interest in finding a "third way" between "economical but polluting" products and "clean though expensive" ones. Looking for this compatibility – which is not a simple trade-off between the two extreme solutions – they perform exactly the cognitive and strategic functions which are typical of policy entrepreneurs in the environmental field.

A second type of resource which can be activated by policy entrepreneurs are *personal links*. Apart from the formal responsibilities attributed to different policy-makers, they have personal histories during which they have connected to a range of people and organizations with whom they are informally in touch. These contacts may date back to their professional formation, to the university period, or to some other experience made in their youth (belonging to the same association or simply attending similar events). Since one of the features of policy entrepreneurs is the capacity to build transversal alliances, the possibility of relying upon solid personal links is a useful resource. These ties allow a quick circulation of information and innovation and, most of all, are based upon a degree of mutual trust, which favors a cooperative attitude. Hence, policy entrepreneurs develop into brokers between different universes – e.g. environmentalists and academics, politicians and interest groups – translating their special languages and becoming a sort of guarantee for the integrity of the bargaining.

A third type of resource which may prove to be useful for the action of policy entrepreneurs is *expertise* and the selective control over *information*. Though both these elements seem to be peculiar of the activity of experts and academics, the range of policy-makers who nowadays have a conscious access to relevant data and information has increased. Environmental associations have their own think-tanks and research institutes, and often conduct their own monitoring of the state of the environment. Private enterprises, if big enough, produce their own innovations in order to find new production technologies. Each country has its own national research organization explicitly devoted to science and technology which, together with several expert groups, produce the sort of knowledge which public administrations look for. The same could be said for the international level, both with the emulation of policy solutions which have proved their usefulness in other contexts, and with specific supranational research institutes and organizations. Free-floating intellectuals and opinion-makers can also on occasion be the catalysts needed for changing the direction of policy making. Finally, universities tend to produce their own wisdom which, not being necessarily tied to specific problems or needs, may represent alternative ways of looking at concrete problems. In this context, policy entrepreneurs may be unable to compete directly in the academic domain according to the mores of knowledge production, but they can favor the promotion of particular ideas and interpretations in various ways and sponsor suitable projects. They link separate circuits, encouraging the cross-fertilization of ideas and exploit the opportunities which open up within different domains.

Beyond these three major categories of resources we could identify other specific factors which support the action of policy entrepreneurs: some kind of *representative investiture* (through elections or as leaders of well-known associations), *past successes* as mediators in conflicts or in the solution of environmental problems, or an *international legitimacy* may certainly help in the difficult task of becoming a recognized leader in environmental science and technology policy-making.

Above we have discussed the basic features which define the action of environmental policy entrepreneurs towards the field of science and technology policy. A further analytical step is to isolate its constitutive principles in order to generate a typology of the practices we have subsumed under the label policy entrepreneurship. Most policy entrepreneurs will rely on more than one of those practices, and probably have recourse to them all at some point during their careers.

Two analytical dimensions seem to be relevant in this particular context. The first is the type of activity involved, whether it is mainly strategic/organizing aimed at network building or primarily cognitive/discursive, that is aimed at promoting environmental paradigms in the science and technology policy process. The second is whether the policy entrepreneur is aiming at creating and maintaining a credible position, or if her efforts are directed at obtaining and retaining a legitimate position. This distinction in turn influences how networks are built and paradigms promoted. Creating and maintaining a credible position implies inserting one more focus of action into a network or introduce a viewpoint into a discussion without necessarily at the same time contesting other possible foci of action or viewpoints. Quite the contrary, this strategy implies seeking connections with other organizations, alliances with other policy entrepreneurs and conceptual links with other established positions. These can be found within the environmental discourse or outside it as the case may be. These activities lead to the establishment of fluid and diffuse networks in which connections change frequently, actual encounters are relatively infrequent except on special occasions, and many centers can be observed. Active alliance building connecting different centers leads to the promotion of abstract, and therefore tentative, open and changeable paradigms which are suited for creating temporary discursive and/or practical consensus around particular events

Working towards and maintaining a position of legitimacy carries different connotations. As legitimacy as distinct from credibility implies an established position in the political process, the policy entrepreneur must, in order to be successful on this account, attend to two powerful imperatives. First, she must focus her network around herself, and delimit this network from other similar, in order to be able to contest the positions of other actors and control her own following. Establishing some kind of organization is the usual means for achieving this. Second, the entrepreneur must pursue aims and proposals for action which are within the (changeable) limits of the political/administrative system. In our case, the legitimate entrepreneur will be interpreting and managing environmental paradigms and translating them into concrete proposal for science and technology policies, and at the same time, actively opposing those of others. Contested priorities and (political) evaluation of technological alternatives within a given paradigm are typical of these activities.

Consequently, the taxonomy of policy entrepreneurship produces four different cells which combine the proposed dimensions. In the following scheme we outline the intersections and suggest different labels for each type of policy entrepreneur.

	Activity	
	Strategic/organizing level (network building)	Cognitive/discursive level (promoting paradigms)
Credibility	multi-center networks	developing paradigms
Legitimacy	single-center networks	managing paradigms

A typology of the practices of policy entrepreneurship

In any country it is possible to find examples of these four practice categories, although the different experiences vary in scope and relevance. At the same time, the national context in which environmental policy-making takes place affects the combination of policy actors – and even of types of entrepreneurial players – which participate in the decision-making process. In the following we therefore give examples of policy entrepreneurship country by country, and relate them to the national contexts for policy making in the environmental and science and technology fields.

2. The Italian Case: Entrepreneurship and Institutionalization

The activities of the first Italian Ministers for the Environment are examples of the lower-left hand box in the four-field classification above. Their mission was to draw together previously dispersed actors with environmental and ecological agendas, who were traditionally underrepresented at the national level. Once in office, they mostly relied upon personal relationships to overcome the fears and skepticism which accompanied their first interventions. Their most immediate problems were organizational rather than cognitive or discursive, and required strategic/organizing rather than intellectual solutions. Their new positions made them obviously potential centers for previously dispersed or multi-centred networks, and forging them into what can be called an embryonic policy community, within which measures can be discussed and proposals fine-tuned. In this they have largely been successful. The current president of the Italian National Agency for the Environment used to be the leader of a (small but militant) environmental as-sociation strongly committed to the establishment of the agency itself, and more or less the same could be said regarding the current Minister for the Environment, for many years spokesman of the Green parliamentary group: they can still rely upon the personal links established in the previous years, both assuring new strength and visibility to interests normally underrepresented, and having the possibility of an immediate external check of the suggestions which come from their own bureaucracy. This policy community, due to its strong relations with the environmental interest, is however very different from e.g. its counterparts in England (cf. van Zwanenberg 1997).

Similar activities are common at the local level in Italy. Many environmental legislative competencies have been transferred to regional and municipal authorities. The most effective local governments often experiment or import successful solutions autonomously, for example the electronic system for controlling the traffic and monitoring the air quality implemented in Bologna. They interact internationally with eco-networks such as ICLEI (International Council of Local Environmental Initiatives) or with the European Union, introducing from abroad the most promising experiences such as the energy budget applied in Livorno and Rovigo, or Local Agenda21 perspectives in Bologna, Modena and Venice. Bureaucrats and politicians of the local authorities act as paradigm managers not only when they introduce in Italy ideas and measures explored somewhere else, but also because they tend to favor the circulation of the most successful policy answers. The municipality of Ferrara, for instance, has recently organized a Web site with *Legambiente* which is one of the most active Italian environmental associations. At this site they present and diffuse the knowledge of successful environmental projects implemented in various places.

An interesting case of political entrepreneurship of the paradigm management type is the recent restructuring of the solid waste policy of the municipality of Milan, after a severe crisis due to the closure of the main garbage dump near the city, which left uncollected garbage in the streets for around twenty days. With an optimal timing, the councilor responsible for the environment, a former environmentalist with a technical background, managed to introduce a few simple policy measures, mainly regarding recycling , which rapidly increased the percentage of sorted and separately collected of garbage from around 5% to around 30%, among the highest in Europe. At the same time, consensus about this solution was created thanks to the mediation of the leader of the Green party in the regional assembly. Now in opposition, he had earlier been in charge of the local environment agency. Both actors had an expertise in the field due to their professional careers and both were (or used to be) insiders of the administrative machine. They could also rely upon the support of the local environmental groups. In virtue of this, they were able to address both the concerns of the citizenry and those of the bureaucratic organizations involved. They were also both deeply engaged in the effort of tackling the problem they were facing. Thus, though with distinct aims, they both interpreted the demands of varied actors for a solution to the garbage crisis in a similar way, and used the opportunity to introduce a solution which was already known elsewhere and somewhat overdue in their own context.

Even environmentalist associations may act as political entrepreneurs, both as paradigm managers, i.e. by pursuing prospective solutions as opportunities arise, but also and perhaps mainly as paradigm developers. *Legambiente*, for instance, was established only in 1980 but has grown rapidly in influence. It has reproduced and spread in Italy many of the ideas of the international environmentalist movement, and adapted them to the Italian context. Its scientific approach to problems has brought about the establishment of the first professional green research institute in Italy, which acts as a consultant of public administrations and private industries. This institute also monitors change in environmental indicators and produces an annual report which the Ministry is unable to improve upon.

Legambiente has recently taken the initiative in creating a database of environmental crimes, and in connecting the different public authorities (police, public prosecutors, etc.) which investigate them, and which traditionally do not communicate or exchange information with one another. Here a new mode of systematizing information has been combined with a networking role. The organization has also, in other instances, worked on building bridges between trade unions and industrial associations. This active networking, which is rather typical of activities in the upper left-hand box, goes well beyond the traditional aims of environmental groups in Italy, which have been oriented towards swaying public opinion and putting pressure on the authorities, activities belonging to the upper right-hand box.

Although policy entrepreneurs are ubiquitous, not every environmental policy requires or encourages this type of entrepreneurship, and there may even be settings and contexts which are hostile to the emergence of such players. Policy entrepreneurship is particularly relevant in situations which are hard to manage through traditional channels. Intractable controversies, technological ambiguity, antithetic problem definitions, complex forecasts and uncertain outcomes, that is when the risk of stalemate is overhanging, these are the signs that a different approach is needed. In cases of weak institutionalization of the environmental orientation to technology policy problems, such as Italy, entrepreneurship is therefore essential. The ability to demonstrate the potential for convergence and consensus between traditionally antagonist actors, deconstructing their policy narratives and outlining new alternatives, this is the capacity which is specific to policy entrepreneurs (Roe 1994). In a certain sense, they come out when other players fail, when the issues are too complex or there is a situation of emergency. Conversely, when policy follows routine paths, is strongly structured by administrative arrangements, does not elicit public attention or is irrevocably petrified in partisan confrontations there is no need (or there is no place) for entrepreneurship. There may always arise implementation problems even for the smoothest issues, but they aren't normally the kind of puzzles which require the networking abilities or conceptual contributions associated to the concept *policy entrepreneurs*, although what Bardach (1977) referred to as *fixers*, that is policy actors which intervene in the implementation phase adjusting certain elements of the process in order to attain the original goals, may be conceived as a sort of minor policy entrepreneur for less problematic topics.

It can therefore be argued that the more the political and administrative context is institutionalized the narrower is the room for entrepreneurial activities. Typical of a strongly institutionalized context is an efficient environmental department which monitors alternative solutions to policy problems and which *internalizes* all the tensions which arise in the social system, which maintains regular organized contacts with interest groups, runs an integrated system for obtaining information regarding the state of the environment and executing the adopted measures, uses well-established channels of communication, etc. And *vice versa*: the more incomplete, *ad hoc*, recently established and erratic the institutional scenario, the more the degrees of freedom is enjoyed by emerging policy entrepreneurs.

Comparatively, the poor institutionalization of Italian environmental policy – with a Ministry established only a decade ago and which is still understaffed, an Environmental Agency which became operative only in 1997, fragmented competencies between central and local governments and endless problems of information management – paradoxically this represents an optimal scenario for the activity of actors which transcend traditional divisions such as that between politics and administration, and between environmental policy and science and technology policy. At the opposite end, the traditional Swedish system, with its old and well organized nature conservation agency, with its highly structured politics, its corporatism, its effective internalization of novelties, its standardization etc., represents an environment unlikely to produce for policy entrepreneurs in any great numbers.

Growing attention to environmental issues in Italy, probably helped by the influence of the European Union, is currently leading towards a process of institutionalization of environmentalism which in many ways is reminiscent of the earlier history of the Swedish system. Here, however, the formative influences are quite different, in that economic actors are involved from the beginning rather than being primarily objects of monitoring and regulation.

Ronchi, the current Minister for the Environment, in charge since 1996 and member of the Green party, for the first time in the governmental coalition, doesn't enjoy the same degrees of freedom of his forerunners Ruffolo, in charge from 1987 to 1992 (in four different governments), or Ripa Di Meana, Minister in 1992 and 1993 and former EC Commissioner for the Environment. His predecessors acted in a sort of political and policy vacuum, and though in a situation of extreme scarcity of resources, their effort to put together a web of environmentally oriented actors was not strongly contested, and Ruffolo managed to accelerate the evolution of his department while attaining the confidence of environmental associations. Ronchi, however, acts in a much more structured setting, and has to take into account the perspectives, legacies and inertia of multiple internal bureaucracies: paradoxically, the information which Ruffolo succeeded to collect in a few months for the publication of the first Report on the Environment, in spite of the lack of administrative resources, seem to be unavailable now, notwithstanding the new special agencies.

3. Lithuania: Policy Entrepreneurship and Transnational Networking

Policy entrepreneurship is visible in the development of environmental S&T policy programs and institutions in Lithuania long before the start of *Perestroika* as well as during and after the period of national revival in the late 1980s. The early and perhaps somewhat premature institutionalization of environmental S&T policy in Lithuania created a fruitful context for the activity of people who transcend traditional boundaries in the policy-making process and manifest entrepreneurial skills in generating and diffusing new ideas, approaches or entire paradigms, bridging respective networks of relevant actors.

During the former period (1970-80s), institutional restructuring and experimenting was to a large extent catalyzed and led by policy entrepreneurs who most often occupied formal posts within the centralized bureaucratic apparatus. Transformations of the former State Committee of Nature Protection could be one example of the institutional change catalyzed and led by such policy entrepreneurs deploying resources they controlled in virtue their formal position. In the Soviet period, the Nature Protection Committee was primarily in charge of nature conservation/ preservation, but since the early 1980s it has gradually shifted towards regulatory engagement in environmental S&T policy conducting and monitoring state-level programs in industrial environmental management. Those programs were explicitly dominated by the state authorities in spite of the general waning of the command economy. However, a closer look indicates important role of the people whom this paper conceptualizes as policy entrepreneurs (Rinkevicius 1997).

Such actors can be characterized as *managers of new paradigms* in environmental S&T policy. By contrast to other local actors, they possess entrepreneurial skills, but utilize them primarily as *general practitioners* who simultaneously introduce new paradigms in the local context and manage their application locally. Dr. Giniunas, the deputy chairman of the State Nature Protection Committee might be mentioned as representing this type of policy entrepreneur. On the one hand, his formal position in the state bureaucratic apparatus legitimized and allowed to formally act as a network builder developing new programs in low-waste industrial technology. For instance, one of such programs was the State Complex Program 82.22 aimed at systemic reduction and treatment of wastewater and solid waste from machine tools manufacturers and electroplating companies. Part of the program was based on utilization of galvanic sludge and residual oil from variety of enterprises at the Palemonas' ceramics factory. Engagement of a number of academic institutes, state committees, agencies and ministries, industrial enterprises was directed by the top-down order of the Council of Ministers listing everybody's "functions" in the network. Initiated in 1982, this program comprised substantial environment-informed R&D activities assigned to the Lithuanian Academy of Sciences and particular institutes: Lithuanian Machine Tools R&D Institute, Thermal Isolation Research Institute, Kaunas Polytechnic Institute, etc.

Although the activity of particular policy entrepreneurs like Dr. Giniunas was important, this emerging network could hardly be described as a single-centered one, as dominated and catalyzed by a one single leader. This reflects important contributions by other entrepreneurial actors who did not hold formal positions in the bureaucratic domain. The formal legitimacy and single-centrality of policy entrepreneurship is reflected in the endorsement and formal enforcement of this program by the top persons at the Central Committee of the Communist Party and the Council of Ministers. On the other hand, the linkages of formal policy entrepreneurs with other relevant actors, groups and organizations, transgressing institutional boundaries between the policy domains, were also important. These actors acted as gate keepers and liaisons (cf. Rogers 1983) in academic organizations (e.g. Lithuanian Machine Tools R&D Institute), industrial enterprises (e.g. Palemonas' Ceramics factory), particular ministries and governmental agencies, etc. Although formally lead by the policy entrepreneurs like Dr. Giniunas of the State Nature Protection

Committee, this newly emerging network was characterized by extensive informal communication, initiative and devotion by entrepreneurs rooted in the academic, and to a lesser extent, in the economic domain.

The outcome of this and similar efforts was a paradigm shift in environmental S&T policy: the 1960-70s in Lithuania were predominantly an era of nature protection, establishment of nature preserves, parks, etc., whereas 1980s were characterized by a shift of focus in environmental S&T policy to the issues of industrial ecology. In this respect, the formal and informal policy entrepreneurs shaping state programs can be likened to the ones which in our typology are described as those developing new paradigms and approaches to environmental S&T policy. There were also those who were engaged in daily practice within the new paradigm being institutionalized, e.g. coordinating involvement of industrial enterprises into existing networks as well as ensuring accountability and feed-back for the furthering of existing policies. On the one hand, such actors are "general practitioners" within the emerging paradigm rather than its creators. On the other hand, their role needs more than routine repetition of institutionalized practices, it requires certain entrepreneurial commitment and the engagement of new actors rooted in the economic and academic domain. There is very little evidence of efforts to extend networking and catalyzing activities to the civic domain. The main efforts have been concentrated in connecting the actors rooted in the bureaucratic, academic and economic domains.

The institutionalization of environmental authorities in Lithuania provides a number of case stories of policy entrepreneurship shaping environmental S&T policy. Some of this entrepreneurship is reflected in the intra-domain type of policy change. One example is recombining in the bureaucratic domain the separate organizational units in charge of water, air and land protection, and creating a single regulatory body, tackling particular polluters rather than different media of the environment.

This process to a large extent depended on the entrepreneurship of actors formally rooted in the governmental agencies, but also on a new type of policy entrepreneurs who did not have a formal post in the public administration. The new policy entrepreneurs, often coming from the academic institutes, for example, Dr. Andrikis of the Lithuanian Institute of Economics, were developing and defending new approaches to environmental S&T policy and administration. Their entrepreneurial activity can be also viewed as aiming at developing a new paradigm. For example, besides institutionally recombining the control of air, water and soil pollution, their entrepreneurship was aimed at integrating an eco-modernist approach based on "polluter pays" principle and environmental taxation practice into this new body of public environmental administration which later became the Department Environmental Protection. It was instituted as directly reporting to the Parliament thus minimizing susceptibility of environmental authorities to institutional pressures by other ministries dominating in the economic sphere. This can be seen as another policy innovation pursued by those new kinds of policy entrepreneurs. The key actors, as Dr. Andrikis, who catalyzed and shaped such changes were not connected with the bureaucratic domain in any formal ways and differed significantly as compared to their predecessors developing and defending new environmental S&T policy paradigms in the 1980s. However, their solid academic background and status as well as entrepreneurial skills allowed them to bridge and lead networks shaping new institutions and policy reforms.

Moreover, they opened up ways for the younger-generation environmental economists, landscape planners and other entrepreneurial "eco-modernists" to get established at the Department (transformed into the Ministry in 1994) of Environmental Protection. Initially being the general practitioners gradually fostering the new environmental S&T policy approach, these younger actors became important policy entrepreneurs not just defending the new regulatory regime, but also actively developing new networks which lead to new policy initiatives and innovations. This type of policy entrepreneurship involves in particular active networking and brokerage in relation to the Western countries and actors there promoting cleaner production, environmental management systems, eco-labelling, green taxes, ecological investment funds, etc.

One important feature of such policy entrepreneurs is their multi-lingual skills, both in terms of knowing international vocabulary of environmental S&T, economics and policy, and in terms of knowing English. They use such skills not mainly as liaisons who by definition are "individual(s) who interpersonally connects two or more cliques within a system, without himself belonging to either clique" (Rogers 1983:101). By contrast, they act more like the gatekeepers who are "located in a communication structure so as to control the messages flowing through a given communication channel" (*ibid.*).

Due to their formal position at the Ministry of Environmental Protection, they legitimize and build single-centered networks which promulgate certain policy initiatives and (latently) inhibit other ones. Arunas Kundrotas, a young environmental economist, who after Dr. Andrikis was Head of the Economics and Strategy Department at the ministry and later a Deputy Minister of the Environment, is an example of such a policy entrepreneur. From 1993 till 1997 he was a formal liaison for all foreign environmental investment projects in Lithuania and coordinated efforts of donor organizations to support environmental technology adoption in the country. Kundrotas acted, however, not only as a formal liaison, but as an innovative gatekeeper, particularly in the field of wastewater treatment technology transfer projects. During this period, some twenty municipal wastewater treatment projects have been started and developed at particular municipalities and involving Western governments (especially in EU countries), EBRD, The World Bank, and other donor agencies. Wastewater treatment plants were also included as a top-priority in the Lithuanian Public Investment Program.

All such efforts and achievements were inseparable from an entrepreneurial activity of policy-makers, such as Kundrotas. An important aspect is that besides the formal leadership in the network, such actors gain the informal trust and credibility among other actors involved in their network. This credibility in interpersonal relationships becomes explicit after a certain period of time, and is mostly latent during the years of their formal enrollment as officers at the ministry and other public institutions. For instance, Kundrotas, after leaving the post of vice-minister, became an official full-time employed liaison in Lithuania of the Danish Environmental Agency. The latter is funding or facilitating investment in at least 75% of waste water treatment technology transfer projects in Lithuania, i.e. the sphere of environmental S&T policy for which Arunas Kundrotas was directly responsible as a deputy minister of the environmental protection.

Inesis Kishkis could be another example of such a gatekeeper-type policy entrepreneur. He worked for several years as a Head of International Relations of the Ministry of Environmental Protection where he led the Lithuania's preparatory work for the UNCED in Rio with Dr. Andrikis, innovatively managed various programs and projects through which Lithuania joined several important international conventions and launched bilateral collaborative programs, including a number of cleaner technology transfer projects with the Danish EPA and other foreign agencies. Kishkis moved on to work as a full-time local environmental specialist for the World Bank after leaving the ministry. Such career trajectories indicate that those policy entrepreneurs during the period of their formal service at the ministry and other public agencies utilize their entrepreneurial skills and formal status /public legitimacy not only for promoting new paradigms advocated by formal eco-networks such as UNCED or donor organizations transferring environmental technology. They also develop informal links and utilize expertise and contacts in those networks to create new gate-keeper niche and status for themselves in the field of environmental S&T policy after they quit the public administration, and thus lose their formal legitimacy as state policy-makers. In this instance, the process implied above, from credibility to legitimacy, is reversed.

This gatekeeper-type of policy entrepreneurship by people holding posts in the bureaucratic domain differs significantly from the type of entrepreneurship which actors rooted in the academic or economic domain, promoting, for instance, cleaner production and environmental management are pursuing. They do not possess a formal status or occupy recognized positions in the policy-making bureaucratic structures of the state. However, the academic or economic entrepreneurs are seeking to acquire such recognized positions – become policy-makers.

This they do by pursuing ministerial/governmental decisions under which certain programs or organizations (and people behind them) would become formally legitimized as significant for the state environmental S&T policy. Examples of such entrepreneurial activity in this field are attempts by different actors and coalitions to launch governmentally endorsed National Cleaner Production Centers, National Cleaner Production Programs, National Environmental Auditing, Verification and Certification authorities, Environmental Standardization Boards, Environmental Education Committees, etc. One example of such a policy entrepreneurship in Lithuania is the Ecological Engineering Association, affiliated with the Confederation of Industries, seeking authorization as the major environmental auditing institution in Lithuania and attempting to get a special order issued by the Ministry of Environmental Protection to that effect. Another similar example is the Institute of Environmental Engineering at Kaunas University of Technology, seeking the support of the Ministry of Economics to pass through the Lithuanian Government a decision on the establishment of the National Cleaner Production Program based at the Institute.

In both cases, particular policy entrepreneurs, one affiliated with the confederation of industries (economic domain) and the other rooted in the university system (academic domain) are introducing and promoting policy novelties. In both cases a particular policy innovation is being brought in from abroad (EMAS from EU, cleaner production from Norway and Denmark) and adopted to the local culture and institutional circumstances in Lithuania. And in both cases, entrepreneurs pursuing policy innovations and paradigmatic changes in environmental S&T policy are seeking to assure for themselves or their organizations certain formal prerogatives which would place them in the center of particular policy-making segments.

In general, two paths of policy entrepreneurship can be observed. In both particular actors – policy entrepreneurs – are creating, promoting or diffusing particular environmental S&T policy innovations, and in both cases it involves credibility and legitimacy, focused and more diffuse networking. The difference lies in the tendency for the policy entrepreneurs rooted in the bureaucratic domain and having formal legitimacy/ resources to develop a gatekeeping role and a personal credibility which they exploit after they leave the ministry/agency. By contrast, policy entrepreneurs rooted in the economic or academic domain, besides promoting some policy innovations, are seeking to acquire legitimacy for themselves and their organizations through formal endorsement by the state authorities and by respective legal acts. Thus, they seek to create (and often do achieve) a demand for their (supposedly) unique competence in those novel spheres, but also to institutionalize their status as the ones in charge of certain programs or institutions in their respective fields.

An important aspect of such entrepreneurial activity is that opportunities for policy-innovation depend on (governmental) funding, and in the case of Lithuania and other transitional economies, such funding is often provided by the governments of foreign countries. Thus, the steps that local policy entrepreneurs, say promoting cleaner production, are undertaking, often depend on the requirements of their foreign counterparts who are seeking to convince their own governments of the relevance and significance of particular policy initiatives and environmental S&T transfer programs from their respective countries to Lithuania. Thus, their policy entrepreneurship has multiple objectives:

1) to promote the new approaches in environmental S&T policy, for instance, cleaner production (cleaner technology), eco-auditing and management, etc.;

2) to fulfill the expectations of donor counterparts to ensure commitment and approval by the governmental authorities, industrial associations and other relevant actors and organizations in Lithuania.

A similar pattern can with local variations be observed in other countries in Central and Eastern Europe.

4. Sweden: A Drifting Consensus

In Sweden responses to and definitions of environmental problems build on a comparatively long tradition (cf. Jamison et al 1990, Jamison 1997). However, the forms of this institutionalization as well as the previous paradigms of problem definition have increasingly been the subject of revision simultaneously with other areas of governance in Sweden. This both applies to the interests and groups consulted in policy making and entrusted with policy implementation, and to the general thrust of environmental policy which in turn sets the parameters for "green" science and technology policy initiatives. This has created considerable room for policy entrepreneurship outside the traditional domains of environmental policy and science and technology policy.

Generally, the capacity of the Swedish system to internalize quickly any policy initiatives before their eventual critical potential is manifested, so characteristic in the heyday of corporatism, has been eroded through two political mantras, namely the reduction of the public sector, and the reduction of public policy interventions in all spheres of life including the economic. This process which both denotes the demise of corporatist governance and the revival of corporate hegemony has led to a move away from regulative efforts and an increasing reliance on economic mechanisms and decentralized initiatives

As a result, the earlier emphasis on environmental science and the identification of environmental problems, determination of maximum emission limits and the development of instruments for measuring pollution has been replaced by an orientation towards solving the problems already known and defined, through adaptation and redesign of industrial processes, consumer goods, building and construction. Among the initiatives signaling this shift are a SEK 5.6 billion project for "Sustainable Society" most of which is spent on municipal sustainability projects focusing among other areas public housing and infrastructure (cf. Eriksson 1996), the establishment of a new research foundation for "solution oriented" environmental research, the establishment of an environmental economics and management, new courses and programs in environmental management and technology at several colleges and universities, and the "Green Shopping" campaign of the main environmental movement organization in the country. Hence policy entrepreneurship today is in many ways focused on interpreting this shift, identifying the opportunities it creates and acting on them.

Various initiatives in this sphere of innovation activity (and the idea of environmentally sustainable technologies itself, of course) are often politically motivated, either in the sense that they are responses to widely recognized problems and concerns raised by members of the public, or in the sense of being creations of the dynamics of the political system as such. Further, politicians tend to be forced by the vicissitudes of electoral politics, the need for legitimacy, etc. to relate to all policy domains more or less simultaneously, or at the very least, relate to actors from within each domain, who purvey the policy viewpoints and concerns characteristic of each. This creates important opportunities for both former activists and bureaucrats for entrepreneurial action. The latter implement, but also – at least in Sweden – often prepare environmental policy to the point of actually making it (Hedrén 1994). However, new opportunities for academics and business people have opened up as well.

An example of the peculiar role of politics is MISTRA (Foundation for Strategic Environmental Research), created by the latest right wing government (or bourgeois government, as it is called in Sweden) in order to finance strategic research on solutions to environmental problems. The form is cooperative ventures between universities and companies or branch organizations. In the initial set-up, this research foundation was to be directed by representatives from the research community and corporate actors, but the current government decided to assume the power of appointing the board. Although civic organizations, which in the Swedish environmental policy making context is more or less equal with the dominating *Naturskyddsföreningen* (SNF, literally Nature Protection Society) are ostensibly to be consulted, their involvement is in fact low.

In this case the political influence is focused on providing earmarked funds for a particular type of research constellations. MISTRA was established simultaneously with a number of other research foundations, which are concerned with medical and technical research. A cultural science foundation was also created which is administered by the Central Bank Bicentenary Fund, another foundation which is an established operator in that field of research and in effect saw its funding increased significantly. Lastly, a special foundation which funds the International Institute for Industrial Environmental Economics (IIIEE) at the University of Lund was established at the same time. This Institute is mainly dedicated to spreading the message about clean production, that is, a new design principle.

In the case of IIIEE an influential broker/entrepreneur can readily be identified, who now is director of the institute, and earlier worked within the government industry support system. The MISTRA case does not appear equally clear cut. Its current director is former Chief of Research at the Environmental Protection Agency, which under his tenure ran a large number of laboratories which studied pollution problems, established maximum emission levels and the like. This may suggest a similar continuity in research priorities as can be detected in the foundation creation operation as a whole. The whole exercise was aimed at creating a better administrative framework for the implementation of industry/academic cooperation in Swedish research. Many – particularly political actors – have long seen this as a panacea to the current woes of the Swedish industrial system which includes a large if shrinking component of sunset industries. However, such cooperation has only materialized slowly within earlier forms of funding administration.

This assessment, however, must be qualified by two observations. First, MISTRA is intended to support mainly solution oriented pre-commercial research, rather than research that identifies/constructs new environmental problems. Second, the foundation itself is cast in the role of a broker which is to bring together actors from industry, academia, public interest organizations and government agencies, using the classical instruments of science and technology policy, research grants and technology development subsidies. Initially, however, scientists have mainly seen the new foundation as yet another funding opportunity. Consequently, the foundation has mainly been offered projects which were already in the planning stages, and these projects were also for obvious reasons better developed at first than those constructed in response to the specific mandate of the foundation.

Considering policy entrepreneurship, incorporation of entrepreneurs is a salient issue in the Swedish context and this issue can be illustrated by the current role of SNF. The phenomenon which has come to be called corporatism, that is successful regulation and encapsulation of the labor/capital conflict, has played a large role in the Swedish political/administrative system. Although the main locus of corporatist efforts has been the conflict between labor and capital, it has formed administrative traditions in general, and this has led to a very strong position for an organizationally united social democracy which was in government continuously between 1934 and 1973, vis-à-vis a split liberal/right wing opposition. This period has left its indelible mark on the Swedish political/administrative system, but also, and perhaps more relevant to the immediate concerns of this paper, formed political discourse and the negotiation of conflicts in particular ways.

Among the important events which cast doubt on the continued relevance of corporatist resolution in the sixties and seventies was the emergence of the new environmental movement, as well as the development of radical feminism, a critical student movement and a period of labor militancy to which strike statistics bear eloquent witness (Sverrisson 1989). Although these movements were related in varied ways and all of them shared a beginning on "the outside" each followed its specific trajectory.

At this point it is useful to relate to Eyerman and Jamison's discussion of the "typical" life cycle of a social movement (Eyerman and Jamison 1991). Emerging on the outside, the movement introduces emergent issues which have not up to then been accorded high priority on the political agenda. However, as time passes, movements in general, and this applies to the environmental movement in particular, create a discursive space within which they construct, as it were, new forms of knowledge, suited to the practices of the movement and suggesting a set of different practices for society as a whole.

The high point of this process insofar as the environmental movement in Sweden is concerned was probably the so-called environmental election in 1988 (Bennulf 1994). From this point on, we can say that the environmental movement entered a new stage. It acquired a position within the established political culture, environmental issues were accepted as legitimate and important political concerns, environmentally oriented science and technology gained roots at the universities and simultaneously, the radical voices calling for major redirection of social development (whom a corporate leader recently termed "environmental fundamentalists") were relegated to the political periphery. From being a credible actor it became a legitimate one.

This state of affairs is strongly reflected in the practice of SNF, which in Sweden absorbed significant parts of the new environmental movement, multiplied its membership during a few years in the middle eighties, and became the main mouthpiece of environmental concerns vis-à-vis the administration. Various proposals are routinely sent by departments and lawmakers to this organization before decisions are made, but the organization or key actors within it are also consulted informally at earlier stages. It is also common that personnel moves from the organization into the state apparatus and back.

This close connection is facilitated by the character of the Ministry of Environment, which is recently established, the personnel of which is fairly young and most of whom have a past as environmental movement sympathizers or as active members, but not necessarily with a high profile in the movement. SNF, in turn, approaches its role in the policy making process pragmatically. It participates in various government committees where its (formal or informal) representatives function as experts (which they often are qualified to do because of their professional competence) rather than as activists. Through its *bra miljöval* (good environmental choice) certification system the organization has also established a day-to-day cooperation with industry. At the same time, the organization also conducts and maintains a publicly critical attitude to both industrial and political interests. In this regard the policy of SNF can be characterized as a "one step ahead" policy. It has for example reacted to government proposals to protect biodiversity with the creation of additional nature reserves in old forests by demanding more of the same, criticizing simultaneously the forest companies for their resistance to ecosystem protection programs in the forest. A similar tactic is employed in the certification activity: as the recommendations of *bra miljöval* become general practice and most products (e.g. detergents) get the certification, stricter criteria are developed as a precondition for keeping the certification; in this way the certification can retain its efficiency as an identification of "environmentally best practice."

This leads naturally to another problematic hinted at above, of *movement vs. organization*. The distinction between movement and organization is particularly important because movements tend to become/take over organizations or fade away. Once movements become organizations, the "iron law of oligarchy" identified by Robert Michels (1948) starts to work: full time professionals, expert organizers, and specialist producers of research, arguments and policies assume leadership.

The environmental movement in Sweden has both created organizations and "taken over" existing ones. The formerly conservation oriented SNF in particular changed its agenda through the influence of the environmental movement and the entrance of new members with an environmentalist rather than a conservationist orientation during the eighties, although its conservationist and "nature-loving" past is still reflected in local activities as well the publications of the organization. Its regular contacts with the bureaucracy and industry have increased the trend towards professionalization, effective campaigns have placed similar demands on the leadership, and it now employs around 50 people in its headquarters in Stockholm and around the country. In this process it has moved from its position as a mouthpiece of a civic environmentalist and activist political culture to become a broker of sorts, maintaining links between the different domains and channeling personnel transfer between them as well, but also provided a platform for critical attitudes and for proposing new interpretations of the reigning consensus.

The connections of SNF to the worlds of academic and practical engineering and science research are mainly on an ad hoc basis, however. Although a large part of their membership is academics and of the leadership as well, as evidenced by the professional qualifications of the members of its steering committee, no systematic policy towards this sphere exists. This can undoubtedly be explained at least in part by the history of environmentalism (as distinct from conservationist ideas and local anti-pollution campaigns) as a science based (societal) criticism, closely related to interpretations of the natural world (ecology, deep ecology, global threats, etc.). In a sense, environmentalism originates as much in the academic domain (if in its peripheral regions) as in the civic sphere. The forms of knowledge underpinning environmental politics and policy making have been more characteristic of the academic form of knowledge, that is science, than of the practically oriented (engineering) form prevalent in the economic/business domain, or the concrete practical forms which constitute what Alan Irwin (1995) has termed citizen science. Hence, if the organization has to work hard to gain the ear of the public and of politicians and convince industrialists and other businessmen of the soundness of its practice and positions the relations with the academic world have been relatively unproblematic and therefore informal and unorganized.

However, the now established legitimacy of the organization and of environmental concerns in general may be eroding the basis of this fairly unproblematic relationship with at least particular types of science (e.g. ecology, biology, organic chemistry and biochemistry). As the stakes get higher critical voices which doubt the scientific soundness of particular positions taken by the organization on different issues have become more numerous, and its dependence on the expertise of persons which happen to be there and their interpretations of the current state of the art in their respective specialties may have to give way for more systematic evaluation of scientific and engineering information.

At the moment, the actor constellation of the eco-modernist project in Sweden can be characterized as a volatile combination of business, academic, civic and political actors each of which are interpreting the emblematic issues of today. Simultaneously, they are anticipating a number of possible futures, and adapting their practices to those perceived as most likely. In this situation, neither heavy handed regulation nor corporatist consensus creation is likely to be adequate, and a number of policy entrepreneurs can and do develop their own initiatives. In addition to those already mentioned, a large number of municipalities have in one way or another benefited from government funding of sustainable projects, and at the local level, networks of businessmen, bureaucrats, environmentalists and, occasionally, academics, are forming around this funding. The relationship of these activities to actual technological practices may appear tenuous at times. In Sweden, after all, environmental policy and science and technology policy have traditionally been negotiated through different networks. However, the "solution orientations", the importance of which is growing, creates strong incentives for a combination of the two, both discursively, which is occurring at the moment, and organizationally, the initial steps of which can be observed in e.g. the activities of MISTRA, SNF and IIIEE. In this process, new forms of policy entrepreneurship will therefore continue to appear.

5. Conclusion

In this chapter we have focused on policy entrepreneurs, the main agenda of which is to bring environmental issues into the process of developing science and technology policy and into the processes in which different technological options are evaluated. It is of particular interest to investigate and compare the role of these entrepreneurs vis-à-vis existing policy actors, the networks that obtain among them and other policy mechanisms. Further, given their potential for enhancing the range of inputs to S&T policy-making and implementation, it is of interest to relate the activities of the entrepreneurs to the potential for democratic policy making and consider how different actors achieve legitimate positions in that context.

Policy entrepreneurs do not necessarily possess formal legitimacy as actors in policy-making, but they are nevertheless often involved in crucial junctures of the decision-making process. They are rarely professional politicians, though they tend to suggest new courses of action for collective actors. They rarely present themselves as external consultants, yet they are certainly experts in their field, and their competence is often different from what is normally found inside the public administration. They even rarely think of themselves as forming a separate category but yet they possess a distinct combination of competencies, which enables them to propose new solutions and create the connections that make them work.

Various features of the policy-making style of a country may affect the potential for policy entrepreneurship. An etatist approach may inhibit it or quickly internalize entrepreneurial initiatives which is a time honored tradition in Sweden. On the other hand, the delegation of responsibilities to the market, as in the "privatization" of environmental policy-making characteristic of the United Kingdom (van Zwanenberg 1997) may force to the fore, as it were, policy entrepreneurs based in the private sector. Inadequate institutionalization creates a wide scope for entrepreneurship, as we saw in the Italian case. When this is coupled to a structural dependence on technology transfer from abroad as in Lithuania the result is a transnational form of poli-cy entrepreneurship, which works differently in many ways than entrepreneurship within national systems of innovation and science and technology policy making.

At the same time, the probability of the rise of unconventional players is not directly related to their actual effectiveness. Opposing weak institutions maybe easier than facing potent ones, but substituting them or reversing the prevalent trend of policy-making in one country is a completely different matter. Where the existing dynamics permit an independent role to single citizens or policy-makers, the magnitude of their influence is probably modest. Conversely, where policy entrepreneurs succeed to by-pass the obstacles and rigidities mounted by powerful institutions, then their influence may be incisive and far-reaching.

For this reason, the impact of innovative solutions may turn to be constrained by their success or, better, the diffusion of entrepreneurship has the corollary of limiting the scope of the entrepreneurs' activities. In Italy as well as Sweden, there have been quite a few local experiences of by-passing the traditional division of labor in the field of environmental policy-making – with environmental associations taking public responsibilities, experts formulating goals, politicians following a policy measures down to its implementation and bureaucrats soliciting the inclusion of the citizenry in decision making – but they have been scattered examples, and their cumulative effect has not been that of reforming the environmental policy arena itself. The entrepreneurial Minister Ruffolo, already quoted for his activism and for having established lively webs of actors in the "infancy" of the Italian environmental policy, could not stabilize the reforms he wanted to introduce, whereas some sectoral innovations brought up by the current Minister, which we have seen enjoying little freedom of action, are set to have a larger impact. Similar trends can be observed in Sweden: The interpretative primacy of governments gives them great power, but only so long as they attend to the parameters implied by industrial configuration of the country. In this context, policy entrepreneurship tends to be limited to interpretation of the changes initiated by governments and decentralized paradigm management.

References

Bardach E. (1977) *The Implementation Game*, MIT Press.
Bennulf, M. (1995) *Miljöopinionen i Sverige* (Göteborg Studies in Politics no. 30) Universitetsförlaget Dialogos.
Bourdieu, P. (1980) *Le sens pratique,* Éd. du Minuit.
Burt, R.S. (1992) *Structural Holes: The Social Structure of Competition,* Harvard University Press.
Dente B. (1994) 'Sviluppo sostenibile e democrazia sono compatibili?', in *Queste Instituzioni*, 22, n. 97: 89-93.
Elzinga, A. and Jamison, A. (1995) 'Changing Policy Agendas in Science and Technology', in S. Jasanoff et al. (eds.), *Handbook of Science and Technology Studies*, Sage.
Eriksson, O. (1996) *Bygg om Sverige till bärkraft,* ABF.
Eyerman, R. and A. Jamison (1991) *Social Movements: A Cognitive Approach,* Polity.
Hajer, M. (1995) *The Politics of Environmental Discourse,* Oxford University Press.
Hedrén, J. (1994) *Miljöpolitikens natur* (Linköping studies in arts and science no.110).
Irwin, A. (1995) *Citizen Science*, Routledge.

Jamison, A., et al (1990) *The Making of the New Environmental Consciousness: A Comparative Study of the Environmental Movements in Sweden, Denmark and the Netherlands*, Edinburgh University Press.

Jamison, A. (1996) 'The Shaping of the Global Environmental Agenda: the Role of Non-governmental Organizations,' in S. Lash et al. (eds.), *Risk, Environment, Modernity*, Sage.

Jamison, A. (1997) 'Sweden: The Dilemmas of Polarization,' in A. Jamison and P. Østby (eds.), *Public Participation and Sustainable Development: Comparing European Experiences,* Aalborg University Press.

Kingdon, J. (1984) *Agendas, Alternatives and Public Policies*, Little Brown.

Majone, G. (1989) *Evidence, Argument and Persuasion in the Policy Process*, Yale University Press.

Michels, R. (1948) *Political Parties: A Sociological Study of the Oligarchic Tendencies of Modern Democracy*, Dover.

Milward B.H. (1980) 'Policy entrepreneurship and bureaucratic demand creation,' in H. Ingram and D. Mann, *Why Policies Succeed or Fail*, Sage.

Rinkevicius, L. (1997) 'Lithuania: Environmental Awareness and National Independence,' in Jamison and P. Østby (eds.), *Public Participation and Sustainable Development: Comparing European Experiences*, Aalborg University Press.

Roe E. (1994) *Narrative Policy Analysis: Theory and Practice*, Duke University Press.

Rogers, E. M. (1983) *Diffusion of Innovations* (3. ed.), Free Press.

Schon D. (1971) *Beyond the Stable State*, Random House.

Svenning, M. (1996) *Miljökriget: miljöarenan och politikens möjligheter att styra vår miljö,* Lorentz.

Sverrisson, A. (1989) *Försvarsvilja, opinionsklimat och massmedier: en explorativ undersökning,* SPF.

Wilson J.Q. (ed.) (1980) *The Politics of Regulation*, Basic Books.

van Zwanenberg, P. (1997) 'The British National Experience' in A. Jamison and P. Østby (eds.), *Public Participation and Sustainable Development: Comparing European Experiences,* Aalborg University Press.

Chapter Three

Representing the Public: New Roles for Environmental Organizations

by Kees Dekker, Mario Diani, Andrew Jamison and Lise Kvande

1. Introduction

This chapter focuses on the transformations that have been taking place within environmental organizations in relation to science and technology policy. As has been noted by many observers, the social movement organizations that were so prominent in the 1970s, when environmentalism represented for many an emerging alternative mode of knowledge production, based on an ecological world-view and democratic organizational forms, have given way in the 1990s to institutionalized and highly professionalized "non-governmental organizations" (cf Jamison 1996, Donati 1996, Jordan and Maloney 1997). Among other things, these NGOs provide professional expertise for research and public education programs, lobby for legislative and policy reforms, and carry out international development assistance projects.

In this chapter, we examine the ways in which the confrontational strategies of the past have tended to be replaced by more conventional, and consensual, forms of activity on the part of environmental organizations. In many European countries, representatives of major environmental groups are granted access to formal policy bodies and procedures, such as hearings or ministerial committees. Provision of expertise and advice to state agencies and private companies, either through formal or informal channels, has also become increasingly important. In programs of eco-labelling and sustainable transport, for example, environmental organizations often play an important advisory role, as they do in many local Agenda 21 projects. In order to be successfully conducted, these activities require respectability on the part of environmental groups, and a more professional mode of operation.

This process has been characterized in terms of a transition from "participatory protest organizations" to "public interest lobbies" (Diani 1997);

voluntary activists have been largely replaced by professionals, at least in the incumbency of key roles within environmental NGOs. Consistently with this change, environmental groups seem to secure most of their resources through mass advertising, direct mailing, etc. rather than through their activists' work in the local community; direct action and protest activities, often of a confrontational type, which were so popular among political ecology and anti-nuclear campaigns of the late 1970s – early 1980s, seem to have largely given way to conventional lobbying techniques.

However, this transformation from oppositional movements to heterogeneous clusters of established non-governmental organizations has differed from country to country, and has had different consequences on the mobilization potential on environmental issues. At times, the rise to respectability has tended to weaken the capacity of environmental groups to wage nationally significant political campaigns – as it has been suggested for the Italian case by Donati (1996). In other countries, on the other hand, there has been a resurgence of activism as a kind of reaction to the new roles that the more established NGOs are playing. Particularly in Britain, but also in Sweden, environmental protest has become a part of a new, anti-establishment subpolitical lifestyle, as activists reinvent, in the opposition to highway building and animal experimentation, the personal politics that were so central to the protest movements of the 1960s and 1970s. Evidence of the persistence of the capacity of action at the grassroots may be found in Ireland (Mullally 1995), the UK (Szerszynski 1995), and Germany (Brand 1995).

Other times, the demise of participatory and protest-oriented organizations has led to organizational forms which cannot be assimilated to the public interest lobby model. A peculiar type of "professional protest organization" has developed (Diani 1997), where professionalization and the emergence of a clear cut distinction between professional activists and organization members, whose involvement is largely restricted to the payment of membership fees, combine with the persistent adoption of confrontational protest tactics. Greenpeace represents the closest approximation to this style of action, even allowing for differences among its different national branches.

Yet other times, organizations have floated along the continuum from participatory protest groups to "participatory pressure groups". In those cases, the gradual dismissal of protest has not been matched by a similar lack of interest in the active involvement of one group's rank-and-file members. Participatory structures have remained in place – eg in the form of local branches of nationally based organizations – and have kept attracting direct contribution from members and sympathizers. Rather than protest activities, members' participation has been mostly aimed at voluntary work and in support of ordinary pressure activities such as membership mail campaigns or personal contacts with local politicians. It should also be noted, however, that transformation has sometimes taken an opposite path, from non-protest to protest-oriented styles. Local branches of groups like WWF have for instance shown increasing availability to get involved also in protest activities, along with more traditional styles of campaigning (Diani 1995).

In this chapter, we indicate how these processes are, to a large extent, dependent on the ways in which public participation has been organized, and more specifically, on the relative openness and transparency on the part of state and corporate actors. In the social movement literature, this is referred to as the "political opportunity structures" that affect particular outcomes (cf Tarrow 1994). We also try to indicate how the present situation is rooted in the past, ie, how the behavior of non-governmental organizations is based on the somewhat different histories of environmentalism in each particular country, and the forms that activism has taken. In Sweden, for instance, where the debates over nuclear energy were highly politicized in the 1970s, and led to a deep polarization in the political culture, environmental NGOs have been given new, but highly circumscribed, roles to play in the new programs of sustainable development. The more radical local activism that is to be found in Sweden can be seen in part as a reaction to the relatively closed opportunity structures, but also to the relatively strong incorporation pressures that have afflicted Swedish environmental activism from the beginning. In Denmark and Norway, on the other hand, NGOs are able to play a much more variegated set of roles, in large measure because of the comparative effectiveness with which environmental movements in those countries were able to mobilize a broad opposition in the 1970s. By stimulating new industrial branches (eg wind energy in Denmark) and encouraging new policy doctrines (eg sustainable development in Norway), the movements in both countries have shown their value to the political establishment, and have thus been given more responsibility than in Sweden for the implementation of the new policies. In what follows, we contrast the experiences in Sweden and Norway, and briefly conclude by looking outward to the situation in some of the other PESTO countries.

2. The Swedish Corporatist Regime

The political system in Sweden is described by political scientists as corporate, or corporatist – "a situation in which interest organizations participate in the formation and the execution of public policy" (Öberg 1994, 22). It is often claimed that interest organizations in Sweden have become links between the state and the civil society in a much more significant way than in other countries. In a corporate system the state does not relate directly to the individual citizens, but rather indirectly, through interest organizations. In order to be able to influence politics and participate in the democratic process, citizens are forced to join an interest organization. Certain organizations are given representation in official bodies and official recognition as negotiation partners for the state agencies. In the Swedish corporate system, the parliament delegates substantial power to the corporate structures, of which it is the interest organizations representing the "labour market partners" – the labour union federations and the employers' associations – which are the most influential. According to Bo Rothstein (1992), an important reason why Sweden to a larger extent than other countries has such a strong corporate system is that in the 1930s – what he terms the "formative moment" – the labour unions were given the right to administer unemployment insurance, which gave them a central role in the making of the Swedish welfare state. Öberg, however, argues that corporatism in Sweden has become problematic, for, in recent years, the strong organizations have come to be seen by many as furthering their own (private) interests to the detriment of society at large (Öberg 1994).

Characteristic for the corporate system in Sweden has been the institutional fusion between the formulation and the implementation of policy. Representatives of various interest organizations are delegated policy making authority by the parliament, most significantly perhaps by participating in the governing boards of the different state agencies. This delegation of power is often legitimated by referring to the need for flexibility and adjustment to particular situations (cf Petersson and Söderlind 1993). In such a way, formal interpretation of the law is replaced by a continuous corporate balancing of interests.

According to Rothstein, the intensive contacts between the state agencies and representatives of interest organizations indicate that the state agencies are anything but the neutral instrumental bodies that they often are thought to be:

> How the public administration functions is not only a question of rationality and efficiency. Instead it can have a decisive impact on the

legitimacy of the political system. The contacts that citizens have with the administrative agencies are by far more numerous than the contacts with the political parties and the members of parliament (Rothstein 1992, 71).

Rothstein contends that "administrative agencies [not only] play an important role when politicians have decided what they want to realize [but that] what politicians want to realize is to a large extent steered by the administrative conditions" (ibid., 345). State bureaucratic agencies in Sweden have substantial power, and, compared to other countries, the state agencies in Sweden are relatively independent from the government. The agencies have a formal responsibility to the government as a whole, and not to any one ministry or minister.

As indicated, corporatism in Sweden means that various interest organizations participate in boards of the administrative agencies. But, in 1992, when the conservative-liberal government was in office, the Swedish Employers Confederation (*Sveriges Arbetsgivarföreningen*, SAF) decided to withdraw most of its representatives from the agency boards. SAF argued that a distinction between public and private interests had to be made in order to avoid loyalty conflicts between the interests of the agencies and the interests of the organization. As a consequence of SAF's retreat, the Swedish parliament decided to abolish general interest representation also for the trade union organizations. A new investigative commission was initiated, with the directive that interest representation should primarily take place at the consultative level, and no longer at the management level.

Another characteristic feature of the Swedish corporate model is the role of the state investigative commissions (cf Lundqvist 1996). Many decisions, particularly when it comes to comprehensive reforms and major policy initiatives, have been preceded by public investigations. One reason for the important role of investigative commissions is that the Swedish ministries are relatively small. Instead of expanding the ministries with large investigative and research resources, temporary bodies have been created. What was earlier regarded as a positive characteristic of the Swedish political culture – its quest for compromise and consensus – has now come under attack, however, due to the lack of transparency and openness in the policy-making system (Petersson and Söderlind 1993).

Recent state investigations have become more limited, and their jurisdiction has come to be more precisely formulated. According to Petersson and Söderlind, however, this more limited role has created problems in Swedish policy making. The development of political strategies for the medium and long term suffers, because the investigations have become more focused on short-term problems. While in other countries more comprehensive attempts to formulate long-term programs and visions are undertaken by independent research institutes (often within the academic and civic domains) such bodies, or relatively independent "think tanks", hardly exist in Sweden. Interestingly enough, however, a number of think tanks have developed in recent years, connected to the employers' federation, SAF, which has also become an active publisher of pro-business, neo-liberal literature of various kinds – including anti-environmentalist books and pamphlets (cf Kraft 1997). In Sweden, there has been a long tradition of study organizations, research centers, and various publishing activities associated with the labor union federations, and the SAF initiatives can be seen as attempts to match the efforts of the labor movement. Another interesting recent development in Sweden is the growing influence of private corporations over the universities, both in terms of funding research and new professorial chairs, but also in establishing closer physical links between corporate research activities and the universities. This is particularly noticeable at the many smaller universities that are being established as a part of a more general regional development effort. The new university in Malmö is a particularly significant example of this new trend, which links industrial renewal directly to higher education and research and development policy.

In his discussion of the corporate character of policy making, Rothstein describes the historical relations between the state and the Swedish Society for Nature Conservation (*Svenska Naturskyddsföreningen*, SNF). SNF was founded in 1909, in part by academics critical of the Royal Academy of Science, which since the 18th century, had been the main organization providing expertise on conservation and nature protection. Like its counterparts in other countries, SNF argued for stronger conservation laws and for the establishment of nature reserves and parks. But in Sweden, SNF also became a kind of semi-official body. From the 1920s, SNF got a small subsidy from the state and was regularly asked by the government to take part in investigative commissions and policy deliberations. In the 1950s, SNF was given a more official status, by being represented in the management of the Nature Protection Board, and there were even proposals to transform the organization into a formal state agency. It was decided in the parliament that SNF would replace the Academy of Science in the implementation of nature protection laws. In addition to receiving more resources from the state, the organization was reformed in order to carry out the new tasks, and state officials became members of SNF's board. In this way the boundaries between the state and the civil society were blurred. SNF was, for instance, involved in the deliberations in 1962 to decide where the water power agency Vattenfall should build its (hydroelectric) power installations.

When the Environmental Protection Agency (*Naturvårdsverket*) was established in 1967 this semi-official status of SNF was largely eliminated, and the association was given a primarily educational role in what was to become a new environmental policy sector: arranging conferences, producing publications, organizing study circles and popular education activities (with state support). Without much discussion in the parliament, the government appointed members from the labour market partners to the new EPA board, rather than representatives from environmental organizations, such as SNF. As a result the EPA came to cooperate closely with industry, rather than with the "old" environmental movement. First in 1988, SNF would be represented on the board of the EPA.

Although SNF's semi-official status disappeared in the 1960s, contacts between SNF and the government, and the EPA, have not been insignificant through the years, but they have primarily been personal and informal. It is not unusual, for example, that employees of SNF move to the Ministry of the Environment, and vice versa. According to a representative of the Ministry, a couple of current employees previously worked at SNF, and, according to a representative of SNF, several staff members of SNF have earlier worked for the Ministry. The biologist Stefan Edman, now engaged as an adviser and speech-writer for prime minister Göran Persson, is a former vice chairperson of SNF and has been a member of SNF's expert committee for biodiversity. And the member of the Center party Gunnel Hedman, who represents SNF in the Delegation for the Promotion of Environmentally Adapted Technology, has previously been engaged at the Cabinet Office, when the Center party leader, Olof Johansson was minister of the environment (from 1991 to 1994). It is important to keep these historical experiences in mind, as we examine more closely the current state of affairs.

3. Environmentalism in Sweden

The corporatist political culture has left a strong imprint on the way in which environmentalism has developed in Sweden (Jamison et al. 1990). The hegemony of the Swedish social democratic party has also been important. Although environmentalism has been a very significant political force in Sweden, it has been difficult for an autonomous environmental movement to develop:

> (In the 1960s) environmentalism in its initial development was quickly seized upon as an important issue for the governing social democratic party, which handled the matter in the way it knew best: the formation of a new administrative body, research committees, and investigative commissions. Environmentalism was transformed into a legal and technical matter, a process of measuring, mediating, and balancing the various interests involved. At the local or movement level, environmental problems became a source for revitalising the party organization, a new issue of study circles and debate and a new topic for experts to specialize in (Ibid, 194).

In discussing the development of environmentalism in Sweden, we can distinguish five phases:

1) A period of awakening from the late 1950s to about 1967.
2) A brief organizational period from 1968 to 1973, when the new environmentalism took more coherent form in new groups and collective identitites.
3) The period of anti-nuclear debate from 1973 to 1980.
4) The emergence of "green" politics in the 1980s, with the creation of the Green Party and the professionalization of activism.
5) A period of internationalization from the late 1980s to the present, as environmental organizations have increasingly come to focus their attention on the "global environmental agenda" (cf Jamison 1996).

The roots of the postwar environmental movement are to be found among the nature conservation organizations, primarily SNF, and, in particular, in its youth organization, the Field Biologists. Compared to other countries, the emergence of the new environmental movement came earlier in Sweden; already in the 1950s, young nature lovers were organizing protest actions against pollution, and when Rachel Carson's *Silent Spring* was published in Swedish in 1963, SNF organized a major conference and began to lobby for stronger environmental protection laws. There was a greater continuity in Sweden between what might be termed the old and the new environmental movements. The period of awakening, in which influential Swedish environmental debaters brought the new concerns to widespread public attention, was effectively brought to a close with the creation of the EPA in 1967, and the establishment of new environmental research programs and judicial arrangements.

As in other countries, however, in the late 1960s and early 1970s a variety of new environmental organizations was created, which took issue not merely with pollution and urban sprawl, but with the technological culture as a whole. In Sweden, however, there was less influence from the new left than in neighboring Denmark and Norway, and the more general societal critique that was articulated by environmentalists in other countries was fairly marginal in Sweden. The environmentalist message was quickly taken over by the parliamentary Center party, and, from the beginning, the new movement organizations had a difficult time establishing an autonomous movement "public space".

During the 1970s, the environmental debate in Sweden was dominated by the growing opposition to nuclear energy. And that opposition was strongly incorporated into the established political culture, both the organs of the state as well as the political parties. Nowhere else in Europe was anti-nuclear sentiment so deeply "parliamentarised" as it was in Sweden. In particular the Center party's identification with the issue was a significant factor in elevating the debate to the top of the national political agenda, especially when the party won the parliamentary election in 1976 largely on the basis of its anti-nuclear position. The debate about nuclear energy, dominating the political landscape during the 1970s, was organized in the form of study circles and information campaigns conducted by all political parties and interest organizations, and eventually culminated in a referendum in 1980. The environmental movement organizations never could capture the initiative, and in the referendum campaign, the strong presence of the Center and Communist parties in the "People's Campaign Against Nuclear Energy" became a source of fragmentation and eventually internal dissension.

During the 1980s, a new kind of professional environmentalism developed, with the coming of Greenpeace to Sweden and the entrance into SNF (and the parliamentary political parties) of many of those who had been involved in the movement organizations of the 1970s. At the end of the 1980s, the Green party, which had been created just after the referendum in 1980, made its entry into the parliament.

As Matthias Gustafsson has written, Swedish environmentalism in the 1990s appears to be highly successful: "Concern for the environment is very widespread indeed, having consequences also for the market and the everyday behaviour of many citizens. Furthermore, state environmental policies officially declare sustainability as the goal for all activities in society" (Gustafsson 1993, 39). Organizations like Greenpeace and SNF, and most recently, WWF, the Swedish branch of the World Wildlife Fund, with their professional, technical and result-oriented approaches, have profited from the increased environmental consciousness, while more democratic organizations have suffered a decline in membership: "[t]he notion of a sustainable development [is] connected to a weakening of the more value-oriented environmental/ alternative organizations. The new strong carriers of an environmental consciousness [are] professional, result-oriented organizations, not explicitly advocating radical changes" (ibid, 37).

According to Lennart Lundqvist, there are more than 50 environmental interest organizations in Sweden (Lundqvist 1996). About half of them are active in eco-technology, eco-information and eco-production, and a quarter consist of associations of environmental professionals. Lundqvist classifies 17 organizations as environmental "movement" organizations. The three largest, SNF, Greenpeace and WWF, together have about 450,000 members (or some 5 % of the Swedish population). The smaller movement organizations with a more active profile have only about 20,000 members altogether. With its 185,000 members, SNF is the largest environmental organization in Sweden. In 1995, 79 persons were employed – full-time or part-time – at the office in Stockholm. In Sweden, SNF is the most powerful environmental NGO, and has many contacts with the state. The main projects of SNF are currently biodiversity, sustainable Sweden and environmentally-friendly consumption. Greenpeace Sweden has about 100,000 supporting members, and was carrying out mainly three campaigns in 1996: a campaign against nuclear energy, a campaign against genetic manipulation and a campaign against environmental poisons. A recent campaign is for the maintenance of a primeval forest in the north of Sweden. At Greenpeace's office in Stockholm, 18 persons are employed. WWF, the World Wildlife Fund, has about 150,000 members in Sweden, but only one third of its budget is spent on projects in Sweden.

Besides these established environmental and nature protection NGOs, there are various smaller or new organizations, which play a role in environmental policy in Sweden. The Field Biologists (*Fältbiologerna*) is an organization with about 10,000 members, and has mainly young people and students as members. Previously, the Field Biologists were a youth organization of SNF. However, after conflicts with its mother organization, the Field Biologists became rather independent (Sjöberg 1988). Traditionally, the Field Biologists organize nature excursions and related activities. Also, they launch environmental campaigns, although the Field Biologists are not a typical environmental NGO. Current projects, among others, are an energy project and a primeval forest project.

The Environmental Federation Friends of the Earth (*Miljöförbundet Jordens Vänner*), has, compared with Friends of the Earth in other countries, few members: in 1995 they had about 2300 members. As a result of lack of personal and financial resources, activities are often on an ad hoc basis, for instance when protesting against infrastructure projects, or writing comments on policy proposals.

New environmental organizations NGOs include the network q2000 and The Natural Step. Both NGOs have come into being with the aim to contribute to a sustainable development, following the ideas of the World Commission on Environment and Development (the so-called Brundtland commission). The network q2000 has been created after the UNCED conference in 1992, and engages youth and students in trying to realize a sustainable development, by way of influencing municipalities, companies, the government and the parliament. q2000 aims at the realization of Agenda 21 on the local, national and international level, by way of "innovative and pioneering ideas". Currently, the network consists of about 700 persons. q2000 has worked very much with the local Agenda 21 projects in Sweden, and is involved in evaluation activities. The network works on a project basis.

The foundation The Natural Step (*Det Naturliga Steget*, DNS) has a rather unique position as an environmental NGO. DNS does not try to influence policy, but rather attempts to stimulate companies in developing environmental strategies. DNS uses so-called "consensus documents" which are based on four lifecycle principles. DNS urges companies to accept the validity of these principles, and, as a consequence, to base their environmental strategies on them (Robert 1994). DNS managed with this approach to convince companies like Macdonalds, Electrolux and IKEA of the necessity of a long-term environmental strategy. While it is no novelty that environmental NGOs use scientific knowledge for their activities, DNS has been extremely successful with its approach. However, DNS is also criticized for the way it uses science. According to the physicist Tor Ragnar Gerholm, DNS's success among companies has given legitimation to what are actually non-scientific beliefs (Gerholm 1996). DNS founding principles, following Gerholm, can almost be regarded as an expression of "ecological fundamentalism" rather than science. They have little to do with modern science. Gerholm argues that the four principles violate the laws of thermodynamics.

DNS has organized a number of different professional networks among researchers, engineers, farmers, and business consultants, which have produced consensus documents on various topics. DNS also is establishing networks among lawyers, nurses, secretaries and other occupational groups.

4. NGOs and Policy-Making

To what extent do the old and new environmental organizations participate in environmental science and technology policy making?

According to Lundqvist, opportunities for environmental NGOs in Sweden to participate are limited in that interest representation still characterizes environmental policy in Sweden. In spite of the new policy to appoint members of commissions and agency boards solely on the basis of their personal qualifications, the boards and councils of state agencies which deal with environmental questions, in the view of Lundqvist, still consist of an "iron triangle" of environmental bureaucrats, representatives for industry and experts from the research community ("Miljörörelsen utestängs," *Dagens Nyheter*, February 9 1996).

Of a total of 116 members in these boards and councils, only one represents an environmental NGO. Lundqvist argues that it is remarkable that there aren't any qualified independent persons in the environmental NGOs to participate in the boards of the state bodies and agencies. According to Lundqvist, representatives for environmental NGOs are excluded because the image that bureaucrats, industrial representatives and experts have of environmental competence allows no room for alternative ideas about environmental issues. Lundqvist argues that there is reason to examine whether "the environmental policy agenda is set by bureaucratic rather than democratic politics" (Lundqvist 1996, 325).

The central agency in Swedish environmental policy – the EPA – has developed primarily into what Lundqvist terms an "intelligence-gathering unit", developing information and providing inputs into policy making. Between the environmental administrators of the EPA and the environmental research community, between the bureaucratic and academic policy domains, close relations exist. While the role of the EPA is very strong, the Ministry of Environment and Natural Resources on the other hand is a very small body, depending very much on the professionals in bureaucratic agencies.

According to an investigation by Marie Uhrwing, environmental NGOs participate to a certain extent in state investigations and committees, both as permanent members and as engaged experts. Most common is that they give comments on proposals for legislation in the areas of culture, energy, traffic, development aid, environmental protection, and nature protection questions and participate in hearings concerning environmental and nature protection issues. It was the former government which started to organize hearings in order to gather different organizations to discuss actual questions, and the current government has continued to organize these "dramatic" happenings, although on a rather infrequent basis.

According to Uhrwing, SNF and the WWF are the organizations which participate most. Interesting is that several of the NGOs are, according to the study, quite satisfied with their current level of participation, and do not want to participate more (mainly because of a lack of resources). However, two thirds of the organizations Uhrwing examined are willing to participate more, and about half of the organizations indicate that they feel partly excluded from policy making. The organizations which are willing to participate more and feel excluded are mainly very small NGOs. Greenpeace, for instance, doesn't consider more participation in policy making important, for it sees other, more effective, ways to exert influence (Uhrwing 1996).

In its review, the OECD observes that public participation in environmental politics is comparatively intensive (OECD 1996). The OECD indicates that environmental NGOs have a close relationship with the Ministry of Environment, the EPA and various boards, have good access to data and are represented in official consultative committees. Public participation in general, according to the OECD, is a routine part of the democratic process in Sweden: individuals and representatives of organizations can for example participate by correspondence in the work of parliamentary committees and give comments. According to the OECD, however, the legal rights of environmental NGOs are very limited. For instance, they cannot object to a permit once it is granted, or sue administrative bodies for not implementing, or enforcing, the law. In the words of the OECD report, "As the legal system relies to a large extent on government institutions at the central, regional and municipal levels to take care of environmental concerns, the rights of the public and NGOs in monitoring application of laws are very limited. While the role of citizens is very active within hearing and consultation processes, it is almost impossible for them to use the courts to force full implementation of the law" (OECD 1996, 180).

The proposal for a new, integrated, environmental legislation which is currently being developed in Sweden suggests a change in this situation. It proposes to extend the legal rights of environmental NGOs that have been existing for at least three years, and have more than 1000 members. Such NGOs should get "the right to appeal against sentences and decisions on permission, approval and exemption according to the environmental legislation, decisions of authorities not to intervene or to take action on the basis of the legislation as well as decisions concerning measures in case environmental quality norms have been violated" (SOU 1996:103).

SNF is a conservationist organization strongly influenced by environmentalism, but it does not subscribe to any general environmental paradigm or world view. It does not maintain frequent organized contacts with the academic world except in the case of initiatives to conserve biodiversity, which is a high profile concern. However, many members of the organization are academics and they represent it in a variety of committees, mixing in a curious and interesting way their activist role and their expert role. SNF also maintains its own expert committees to prepare campaigns and information material. However, SNF is not very active in areas traditionally understood as science and technology policy. It has, for example, only received one proposal from MISTRA (The Foundation for Strategic Environmental Research) for review. In transport and energy policy SNF has focused on issues such as infrastructure, automobile fuels and nuclear power, to name a few in which its strong influence on public opinion is evident. Through consumer campaigns and its certification bureau SNF also influences both product and process development in a large number of companies. However, most of their work is focused on incremental innovation, and the organization is not involved as such in alternative new technology development projects, although some of its members are. Bo Nilsson, in a comparison of the environmental strategy of the automobile producer Volvo and three environmental NGOs in Sweden – SNF, MJV and the Field Biologists – argues that the position of these environmental NGOs is quite defensive (Nilsson 1995). Nilsson argues that environmental NGOs in Sweden to a large extent have left technology policy and technical development to others. Rather than aiming at a con-structive interaction with Volvo, or trying to influence the development of technical variations, according to Nilsson, the environmental NGOs primarily try to prevent undesirable solutions, and attempt to influence the public opinion.

Concerning formal public participation and Swedish environmental R&D efforts, the picture is not radically different from the situation for environmental policy. Here, too, the corporatist system has left its mark. Large environmental organizations like SNF and the WWF are represented in a few boards. In the new board of MISTRA, for instance, the general secretary of WWF, Monika Stridsman, is represented. SNF's Gunnel Hedman is represented in the recently established Delegation for the Promotion of Environmentally Adapted Technology. Thus, the recommendation of the OECD to invite representatives of the public to participate in the government-private sector partnerships for cleaner technologies has been followed in a way that seems to be common in Sweden: by offering formal representation to the large interest organizations. Striking is that in Sweden mainly organizations that traditionally are concerned with nature conservation are represented in environmental R&D activities, while new organizations and organizations which emphasise the newer environmental concerns for sustainable development are usually not represented.

5. Environmental Organizations in Norway

The environmental voluntary organizations in Norway have, as in Sweden, gone through different phases. Environmental issues were set on the agenda first from the late 1960s, and several organizations were established during the early 1970s. The period from 1970-75 has been called the "golden era" of environmental protection in Norway, because of the increasing interest in and organizing around environmental issues. The Norwegian environmental movement was said to be the strongest in Europe at that time (Berntsen 1994, 157). Also during the 1980s the organizations were quite strong, with an increasing number of members and a new interest in environmental issues in the last part of the decade.

Most voluntary organizations have seen a decrease in membership during the 1990s. This fact has also affected the environmental organizations. While environmental politics in the 1960s and early 1970s primarily was an issue for organizations in opposition to the political system, the situation shifted from the late 1970s onwards; Now there is a substantial public administration on all levels in the society, which has adopted the environmental issue, while the environmental movement organizations have weakened and have problems in finding their function in a society where many of their tasks have become "mainstream policy".

Most of the new organizations are different in several respects to older organizations, both the classical social movement organizations and the alternative life-style organizations from the 1970s. The classical voluntary organizations in Norway are characterized by a democratic structure and pragmatic way of working. The new organizations from the late 1960s and 1970s were more concerned with ecological and other life-style issues, and, as in other countries, there were a number of major confrontations, in Norway, most dramatically in relation to hydroelectric power. There was also in Norway a number of distinctive ideological, or cosmological, elements in the environmental movement of the 1970s that were to have a major significance. On the one hand, there was the eco-philosophy or "deep ecology" of the philosopher Arne Naess, which in the course of the 1970s and 1980s has spread throughout the world and become an important stream of ecological thought. There was also, particularly in relation to the political campaigns opposed to joining the European Union, a school of ecopopulism that emerged in Norway. And finally, there was the influential writer Erik Damman, who formed the organization, the Future in Our Hands, with a unique program combining global solidarity with ecological and life-style issues. This became an extremely popular organization in Norway (even spreading to other countries).

By the early 1980s, however, all three of these movement "streams" had lost much of their political importance, as did much of the rest of the Norwegian environmental movement, in the wake of the struggle at Alta, where activists unsuccessfully tried for many months to stop the building of a hydroelectric dam. From the mid-1980s, in a kind of neo-liberal period, that has also accompanied the rise of Norway as an oil-producing power, there have developed new types of organizations, with characteristics different from the already existing ones. To give a very brief overview of these new (also called neo-liberal) organizations in Norway:

- They seem less ideological, and more concerned with pragmatic solutions and modes of working; this means that most of them cannot be placed within the left-wing tradition.
- They are less concerned with state politics and administration, and are more oriented towards the market.
- Environmental knowledge and expertise seem more important than earlier.
- Many of the new organizations are not based upon democratic principles.
- The voluntary sector is more marked by a competition between the different organizations.
- Media attention seems even more important than earlier.
- There seems to be a development towards more financial support from the state to projects and activities than to the organizational expenses, thus favouring small, flexible and pragmatic organizations.

It should be noted that several of the older organizations still play an important role, and have met the new challenges in different ways. In the following sections, we present some examples of the new environmental organizational landscape in Norway.

Norwegian Forum for Environment and Development (ForUM)
The Ministry of Environment has made a list of those voluntary environmental organizations and their networks with which they have formal and informal connections. There are seven main networks, or umbrella organizations, concerned with environmental issues that are mentioned:

1) Liaison committee for nature conservation issues (SRN, 4 organizations).
2) Joint organization for outdoor life (FRIFO, 12 org.).

3) Liaison committee for preservation of biological diversity (SABIMA, 8 org.).
4) Joint organization for protection of culture (KORG, 13 org.).
5) The Environmental Home Guard (MHV, 16 org.).
6) Action for Neighbourhood and Traffic (ANT, 9 org.).
7) Norwegian Forum for Environment and Development (ForUM, 57 org.).

Through these networks, several voluntary organizations (environmental and others) have an institutional cooperation which also functions as a channel of communication between the organizations to the government. The activities and power of these networks differ from each other, but they can all be seen as an expression of a new interest in state-civil society cooperation. On the other hand, this kind of voluntary work has little to do with mass mobilization, which has been seen as one of the positive forces of the NGOs in democracy. In this way, the new networks can be interpreted as both a cause and effect of the development towards more bureaucratic organizations.

The most comprehensive network, ForUM, is concerned with Norwegian politics on the international arena, especially within the UN system.

ForUM was started in 1992 as a continuation of "The common campaign for the Earth's environment and development" (*Felleskampanjen for jordas miljø og utvikling*, FMU) from 1987; both can be seen as results of the UN's call for public participation in the report, *Our Common Future*, written by a commission headed by the Norwegian prime minsiter, Gro Harlem Brundtland. The Common Campaign was started by persons from the Ministry of Environment and NORAD/Ministry of foreign affairs, as a one-year project, to follow the proposals from the report, which had been published the same year. It came to be continued until 1992, when it was reorganized. The Common Campaign had two purposes:

1) To attract attention to and spread information about the Brundtland report.
2) To find new modes of cooperation between the NGOs across different sectorial interests.

The Common campaign came to be well integrated into the UN process, as organizer of the European NGOs' participation at the UN conference in Bergen, 1990 and as an actor in the preparation for and participation at the UN conference in Rio 1992. This activity was supported, financially and politically, from both the Ministry of Environment and the Ministry of Foreign Affairs.

It was seen as a major task to make environmental issues a part of all sectors, and the Common campaign succeeded in getting organizations from most societal arenas involved (from industry to peace, sports, farming, women's organizations, religious and workers' organizations, etc.). During its first year, the number of organizations participating in the Campaign grew from 37 to 67, and at the end of 1992 there were 106 organizations involved. They represented different and sometimes conflciting interests, but could handle these differences as long as the main effort was not to make political reform, but to raise questions. Until 1991 the two mentioned ministries were centrally placed in the executive committee, but gradually other interested NGO members took over.

This form of organization and cooperation between NGOs and bureaucracy gave the Common campaign an exceptional position in the Norwegian political system. This was also underlined by the fact that the campaign was given resources to finance environmental activities and projects in the various member organizations. Throughout the 6 years, the campaign allocated more than 10 million NOK to their member organizations, given by the government.

After some years it became clear that there was a gap between the campaign's activity and the ambition to raise the popular environmental consciousness in general. The campaign was too hierarchical, due to its role as an umbrella organization and not a "people's organization". The work for international and political cooperation and knowledge development, could not bring out grassroots engagement. The wish for more action and engagement at the local level was a main reason for the establishment of a new network of national voluntary organizations in October 1991: *Miljøheimevernet* (The Environmental Home Guard, MHV). The National campaign and NNV (The Norwegian Society for the Conservation of Nature) took the initiative in this network, which gathered 16 national organizations in order to get their members interested in environmental practice on the local level.

After the 1992 Rio conference, the National campaign was reorganized. ForUM became a kind of continuation of the Common campaign, though the number of member organizations fell from 106 to 58, divided into "contributing" and "affiliated" organizations. As the situation was with FMU, this includes both "pure" environmental organizations, and organizations for peace, solidarity, women's interests etc. But industrial and working organizations are not on the list of members, neither are professional groups of scientists; so the new umbrella organization has a more clearly defined profile of voluntary organizations. The main financial support still comes from the ministries of environment and foreign affairs.

ForUM is organized with limited leadership, but with several ad hoc working groups. There are at present 5 persons in the secretariat, whose activities are information and co-ordination of the working groups' activities. The network has an executive committee elected by the annual meeting. Each member organization, whatever size, has one vote at the annual meeting.

One of ForUM's primary efforts has been to produce policy and research material for UN meetings, where the network is participating in the Norwegian delegation, representing Norwegian NGOs. Here the working groups are a vital link of the network. The members of the working groups are to have professional and/or ideological interest in the issues, and when a group itself lacks competence, it consults other, professional groups or persons. The working groups make reports and other publications, and they prepare documents for the meetings on development and environment in the UN. Persons from the member organizations concerned with the actual issues, are welcome to par-ticipate in one of the groups. At present there are 9 working groups in ForUM. One is working on biological diversity and sustainable agriculture, one on sustainable consumption, one on debt and development etc. The focal issues are in particular the ones discussed internationally, especially in the UN. This makes ForUM different from most environmental organizations, where primarily Norwegian politics and development are the main forces.

Another of ForUM's efforts has been to keep contact with and collect information from international organizations, such as GATT, GEF, ANPED, Earth Council and Third World Network, and distribute information from these to its member organizations. ForUM takes part in several national and international committees, such as the international steering group for environmental NGOs under the CSD, in the Programme council for environment and development in the Royal Norwegian Research Council (NFR-MU), National Commission for UNESCO, The North-South idea group and Norwegian committee for international environmental questions (NIM). From these the last one is seen as the most important, where the Confederation of Norwegian Business and Industry (NHO) as well as the Norwegian workers' union (LO) and other national environmental networks are represented.

The practice of this network can be seen as a variant of the Norwegian (and Nordic) way in which the organizations are given a place in the political system. ForUMs role in the involvement of the NGOs is both to strengthen communication and information between the organizations and between the organizations and the government, in addition to their role as a representative for their member organizations when it comes to national and international policy making.

As a representative for about 60 organizations, ForUM connects organizations with different backgrounds and makes it easy for the MD to co-operate with the NGOs, helping give public participation a formal, administrative legitimacy.

It seems clear that channels for participation, representation and information for Norwegian voluntary organizations on the international level have been opened and strengthened during the ten years the National Campaign/ForUM has existed. The rather close connection to the Norwegian government is considered positive for both sides, even though ForUM's political ambitions also involve exerting some kind of pressure against the government.

From this point of view we now want to take a closer look at one of the member organizations of ForUM, which does not see the traditional channels of political influence as the most fruitful way of dealing with the environmental challenges.

The Future in Our Hands (FIVH): Integrity Intact?

The organization named *Fremtiden i våre hender* (FIVH, The future in our hands) was established in Norway in 1974, two years after Erik Damman published a book with the same title. More than any other environmental NGO, FIVH has a strong ideological profile, which can be seen as a typical phenomenon of the 1970s – but not of the 1990s. Opinion influence is also seen to be the most important task today.

FIVH has, like most of the NGOs, experienced a decline in number of members in recent years, and currently has about 16.000 members organized in some 20 local groups (6-10 of which are considered to be functioning well). Subscriptions amount more than 80% of their income. The public financial support (from the Ministry of environment) amounts to less than 15% of their income. The secretariat is rather small, with 2 to 3 employed persons. Despite these quite limited resources, FIVH has its own research institute with 3-4 persons involved – though the main spokesman rather would like to call it information than research. Further, FIVH has initiated the establishment of NorWatch, an institute for assessing and controlling Norwegian firms' activities in the developing countries.

The ideology of FIVH can be summed up in the term "limits to growth". The third world focus remains central, and a sustainable development is seen to be possible only if the rich world reduces its consumption, and changes its fundamental economic principles, according to FIVH. At the same time, the spokesman points to the dilemma that changing peoples' habits in one way might lead to new, even worse habits for the environment. For example, if people buy more used products, they save money to spend for leisure activities which implies an increased use of transportation. Thus FIVH, though engaging in the Environmental Home Guard (see below), and despite their reputation as "moralists", has pointed to some of the problems of privatizing environmental politics.

Media attention has from the beginning been seen as a main way to influence the public opinion, and the main spokesman in FIVH has the status of a media figure, and is often taking part in public (media) debates on consumption, global responsibility etc. Because of the limited resources, FIVH pays less attention to political decision-making than many of the other NGOs, and they usually don't use their right to comment on proposed legislation. According to the spokesman, they don't find it fruitful to follow this channel because it takes a lot of time and the political gains they might achieve through this are too limited. But they do pay attention to the political system in different ways. Three party leaders, plus other activists of political parties in Norway are among their members. And Gro Harlem Brundtland has said on numerous occasions that FIVH was a source of inspiration for her work with *Our Common Future*.

The way in which FIVH seeks influence in politics, is, primarily through media, and dissemination of their reports and publications. In this way FIVH follows the same path as most of the other NGOs; their strength can be measured through their production and the use of their products within the knowledge-making field. Their reports are of various quality; they have e.g. mapped the political parties' environmental efforts, and they have presented new thoughts on how the use of cars can be reduced. One of their best known projects is to show how working hours can be reduced, a project that has been carried out in cooperation with the research group *Alternativ Framtid* (now ProSus).

In the Norwegian context, FIVH has established a niche as practical visionaries. They are regarded by many as idealist utopians, and are in that sense cut off from serious politics. On the other hand, they are often praised for their original suggestions, and are thus taken into account both by politicians and others. A former conservative prime minister is among those who has expressed satisfaction with having FIVH in the political landscape. With their concern for limits to growth and North-South problems, they represent the society's bad conscience and they are welcome and even respected, as long as they are small and have limited power.

New Environmental Organizations in Norway

It has been noted that new organizations, which have been established since the 1980s, differ from older organizations in that they often use untraditional methods, they are more action-oriented, and they are often not structured as democratic organizations.

Researchers dealing with voluntary environmental organizations in Norway, have put it this way: "Since the 1980s, the number of very centralised, "non-profit" organizations not based on individual membership has grown, and with this growth, the historically dominant organizational model has begun to lose ground." (Selle and Strømsnes 1996). For our purposes, it is interesting to look at how the new organizations are linked to other organizations, to the politicians, and to the market.

Bellona and Greenpeace are the most well-known of the new generation of environmental organizations in Norway. Both were established in the 1980s, they are based on financial support from contributors, but are not de-mocratic organizations. In the following we will briefly discuss Greenpeace's status in Norway, and present the Bellona Foundation.

The Norwegian Greenpeace, established in 1988, is quite small, with about 1.000 supporters. There are only 4 (5) persons engaged. The Norwegian group is dependent on financial support from Greenpeace International. The campaign leader of Greenpeace Norway, Harald Sævareid, has put it this way: "In many parts of the world, Greenpeace is a highly respected organization, and the mass of members are solid. In Norway Greenpeace has a very low number of members, positive mention is seldom, and the impression is that the organization has very little legitimacy and respect in the society." (Sævareid 1996, p 205).

The campaign leader means that their main problem is a lack of credibility primarily due to actions against whaling. In the mass media's presentation of their activities and opinions, there seems to be a common understanding of Greenpeace as an American-style organization, with an extreme view on nature and protecting animals. This contradiction between Greenpeace and Norwegians became particularly clear during Greenpeace actions, followed by a lot of media attention, against Norwegian whaling in the 1990s. A previous leader of Greenpeace Norway, who was fired in 1992 because of his disagreement on the anti-whaling and anti-sealing actions, has expressed his standpoint in this way: "The perspectives and the future with Greenpeace is horrifying. Thousands of people work and engagement is brought to a dead end while the real environmental problems are left untouched." (Økern 1993, p 142). This objection has also characterized the mass media, and seems to be a general Norwegian opinion.

Greenpeace does not participate in the Norwegian formal networks/ umbrella organizations mentioned above, and there seems to be limited contact between Greenpeace and other environmental organizations in Norway. Greenpeace is small in Norway, not because of its non-democratic way of working, but rather because of its criticizm of central elements in Norwegian relations to nature – and, as in Iceland, criticizm of an important economic activity. The case of Bellona shows that, even in Norway, however, new professionalized environmental organizations are taking over from the old movement organizations of the 1970s.

Bellona was founded by two former members of *Natur og ungdom* (NU, Nature and Youth), the Norwegian equivalent of the Swedish Field Biologists, in 1986. Rune Haaland and Frederic Hauge then often appeared in the media, because of their actions against Norwegian industrial firms. During actions against the firm Titania in Jøssingfjorden, they co-operated with the Danish Greenpeace (Dahl 1996). Hauge and Haaland wanted to start a new organization because they found the existing ones too big, inflexible and ineffective to deal with the shifting environmental challenges.

At the outlet Bellona got a lot of media attention and gave an impression of being a new generation of idealistic and activist youths, doing everything to expose how industrialists polluted air and water. But, according to Haaland, they were always interested in dialogue and problem solving. During the 1990s their profile changed. They are no longer concerned with actions to get public attention, although they still investigate e.g. industrial pollution. The shift towards a different way of working, with less media attention, was announced by Frederic Hauge at the end of the 1980s (Sæten 1991).

One of their main tasks has become to assist industry with knowledge and practical solutions, so they now act more like a consultant firm, or an environmental competence center, than a group of rebels – though they don't sell consultant services as such to industry. They define their role to be "a competent expert – but non-governmental organization, co-operating with authorities when necessary."

In March 1997 *Reader's Digest* nominated the leader of Bellona, Frederic Hauge, "The European of the Year". This is a remarkable award for a Norwegian who dropped out of school. His companion Rune Haaland worked as a school teacher before they founded Bellona, and, according to Haaland he supported Hauge for a while. With this background they now manage a foundation with some 30 employees, many of them highly professional, and with offices in Oslo, Brussels, Washington, St. Petersburg and Murmansk.

Norwegian firms and the Ministry of Foreign Affairs are their most important funders. Today there are about 1,500 supporting members of Bellona (not members in a traditional sense, as Bellona is not a democratic organization). But as opposed to Greenpeace, Bellona has access to influential areas of policy and decision making in Norwegian society.

The financial support from the Ministry of Foreign Affairs was 4,4 mill. NOK in 1996. The main financial basis, however, is the firms, which represents more than half of Bellona's income. The way in which the firms have given support is primarily through advertising in Bellona magasin. Their financial problems, however, not least due to the Nikitin case in Russia when the organization had to pay for the release of one of its employees, have forced the organization to seek new ways to secure their income. They want (and get) grants from industry to specific projects, and in exchange, Bellona offers environmental expertise. And cooperation with Bellona gives a kind of legitimacy to a company's environmental image.

As distinct from most environmental organizations, Bellona's activities are mainly related to technological and industrial developments. Development of new technologies, such as the electric car, is seen as the most fruitful way of dealing with environmental problems (Nilsen 1996).

Bellona seeks influence in the business world, and this means going outside of the traditional channels for NGO political influence. Thus Bellona is not represented in any of the environmental NGO networks. On the other hand they have become highly effective in lobbying on a national, but especially on the international arena. According to the project supervisor, they find it easier to get in touch with the EU commission than with the Norwegian government.

> Haaland and Hauge wanted a small and flexible organization to deal with different and changing challenges. According to Haaland, they like roundtables where there are no opposite parts, where it is possible to talk things over and find common solutions. Regarding this, the foundation are now establishing so-called "reference groups" for their various issues. For instance, within their "strategic plan 1996-2006" they have a "programme for cleaner energy" (which is supported by the Norwegian company Statoil with 700.000 NOK).

Bellona has established a reference group on energy, where representatives from energy firms such as Shell, Statoil, BP and Alcatel participate. The group discusses several problems and challenges facing the energy sector now and in the future. As Haaland puts it, an important thing here is that the group members don't think as firm representatives, but as Bellona: The group is discussing challenges and possible directions for a more environmental development on a general level. Thus it is possible to gather these persons around Bellona's "roundtable". Reference groups like this have several purposes: Bellona gets information on what is going on in industry, and they function to establish and maintain personal relations with experts from the companies.

By these activities, it makes sense to call Bellona an entrepreneur, or a broker, in the Norwegian environmental field. They seem to play a role as "glue" between competing companies, and also on an international level through their work for European support to environmental protection in Russia. Bellona has found its niche between the other environmental organizations, by having a rather weak, or low, ideological profile, and a strong practical orientation. Personal relations to both industrial managers and politicians in Norway and abroad are important for getting results – both in terms of financial support for their projects as well as for producing concrete technical solutions in selected companies. Though their reputation as an environmental organization is questioned by other environmental NGOs, their role and success among industrialists is due to the industry's need for practical environmental advice, and their need for legitimacy through connections to an environmental organization such as Bellona.

The Environmental Home Guard (MHV)

A growing problem for environmental organizations in recent years has been to maintain their membership. In this sense, we can talk about a growing gap between the voluntary organizations and the general public. In Norway, however, steps have been taken towards the general public to make them not only interested in environmental issues, but also to get them to change their behavior towards a more sustainable consumption.

"The common campaign for the earth's environment and development" (FMU) established in 1987, was after some years divided in two new umbrella organizations. While ForUM was a continuation upward in the political system, the Environmental Home Guard (*Miljøheimevernet*, MHV) was established to be an organization for the individual citizens in Norway.

The initiative to establish the MHV was taken by The Norwegian Society for the Conservation of Nature (*Norges Naturvernforbund*, NNV) and FMU, and the idea to create an environmental homeguard came from the general secretary of NNV, Dag Hareide in 1990. The project was founded by direct grants from the Ministry of Environment, and was officially started in October 1991. All the political parties in the parliament supported the official financing of the project, which was supposed to be an experiment.

The main strategy of MHV was to engage as many participants as possible in so-called *dugnad* (voluntary communal work). This rhetorical concept points at a traditional way to carry out local duties and activities by collective effort (work sharing) with no money involved. The *dugnad* was meant to include all sorts of people, not only those with a high level of education and commitment that usually has characterised the Norwegian environmental movement. As the leader of the MHV says: "While the environmental movement has traditionally been exclusive, mainly involving people with high level of competence, commitment and involvement, the EHG [MHV] aims at being inclusive" (Endal 1994, p 34).

The increase in participants 78,000 in 1996 shows that they so far have succeeded in being inclusive. But how could this be possible when we already know that the traditional environmental organizations experienced a decline in members and legitimacy in the same period? To find some of the answers, we'll take a closer look at MHV's strategy.

That the MHV is not a traditional organization but a *dugnad*, means they have participants instead of members. Becoming a participant has so far been free of charge. For a lot of people this is important because they already subscribe to several other organizations. Not paying anything means that it is easier to give it a try. Being a participant instead of a member also means that people are not obliged to go to meetings regularly. These advantages do not, however, mean that the participants are indifferent to what they are supporting.

The idea of the MHV is to motivate people to change habits instead of trying to get a few people to change all their bad environmental habits. Instead of paying a subscription, the participants promise to change at least five environmentally harmful everyday habits, that are suggested among thirty others on "the action plan". The habits they choose to change are registered together with name, address and age.

So far it seems like the *dugnad* concept together with regular information is successful as far as to include and engage people. Investigations show that 70-90 % of the population want to do something about the environmental problems, but they don't know what to do. By informing people about concrete things they can do and the effects it will have on the environment if for instance 100,000 people do the same, it motivates each individual to take part in a joint effort. The amount of participants is therefore very important if this inclusive strategy is going to become a success. Through newsletters, the participants are requested to focus on specific extraordinary topics from time to time.

MHV, as an umbrella organization, has formal contact with 15 of Norway's biggest organizations, representing more than 3.8 million members (on the average, each Norwegian is a member of between 4 and 5 organizations). Through these organizations, information channels and networks, combined with specially trained MHV contact persons within each organization, they are able to reach a lot of people with their message. The network of people with totally different backgrounds, but with the practical environmental efforts in common, seems to be the strength of the MHV.

MHV started with offices in three counties. They are now represented in seven in addition to the secretariat and the information office in Oslo. In the long run they aim at being represented in every county. All the 15 supporting organizations behind the MHV, at central and local level, are acquainted with the connection with MHV. It means they can use the MHV as a tool bringing environmental issues and tasks into the organization. MHV also started cooperating with schools, both pupils and teachers, combining the message of the MHV with the schools own education on environmental topics. Not only did the schools, including students/pupils, administration and teachers together try to change as many environmentally bad habits as possible at school (like garbage-handling, lowering the inner temperature etc.). It also resulted in special campaigns where the students tried to engage parents, family and their neighbourhood to become new participants. As expected, these campaigns resulted in a rather fast increase of participants. This expansion was, according to some of the MHV county-leaders, necessary to legitimate a continued official funding of the MHV.

MHV has focused on special environmental projects on the local and individual level. They have produced several workbooks on baby care, techniques for home composting, practical information on waste separation and recycling etc. These workbooks are meant for individual participants as well as for groups of participants like schools, offices and organizations. A lot of local activities are based on these books in cooperation with the public and the local representative of the MHV (usually the county leader).

Conclusions

The field where non-governmental organizations act is rather wide and heterogeneous, and many of the NGOs have little to do with each other both politically and practically. The differences in ideology, ambitions and target groups also reflect different modes of working. Even the few examples of environmental NGOs we have presented in this chapter testify to the persistent variety of organizational forms. While the growing relevance of professionals within environmental organizations is undisputed, and so is their increasing reliance upon lobbying techniques instead of protest actors, one should not conclude that a uni-directional process of organizational change has taken place. The passage from environmental movements dominated by what we have called "participatory protest organizations" to networks of "public interest lobbies" is far from representing the only pattern. Similarly to its more popular counterparts in other countries, Greenpeace branches in the Scandinavian countries appear to come closest to the profile of the "professional protest organization". Possibly the most striking trait of this model is its retention of a protest repertoire, yet detached from its mass basis. Dominant accounts of political protest have stressed its being a particularly attractive option to those social groups, which are not able to affect the political process by traditional means. The recourse to mass action reflects in this view the emphasis on a "logic of numbers" , as a counterbalance to the greater influence enjoyed by more centrally located and/or better resource-endowed actors. In Greenpeace's case, protest – better: its most spectacular component – is disentangled from mass grassroots participation and left with groups of professionals, coordinated by an organization which by comparison to the stereotype of the movement group is definitely not short in resources. Our investigation also suggests that participatory orientations among environmental NGOs still play a key role, the major difference to previous phases being possibly the reduction in confrontational, protest-oriented participation. Among the groups we have discussed, the Environmental Home Guard (MHV) in Norway seems to be the closest to a model which combines participation and institutionalized patterns of action. In this case the label of "participatory pressure group" proposed by Diani (1997) for organizations of a similar type does not seem to be totally appropriate – MHV's focus is far more on cultural transformation and changes in lifestyles rather than on political pressure. What's most important is, though, the recognition that recent transformations in the profile of environmental NGOs have not necessarily led to the rejection of grass-roots participation. We are rather witnessing the more complex coexistence of several organizational patterns.

Many have pointed at the NGOs as important knowledge producing units. These impressions may be discussed from several points of view. We would rather define the NGOs in general as a filter between science/knowledge on the one hand, and the general public, industry and/or authorities on the other. This point of view does not consider their role as a value-free or passive one; the NGOs certainly produce, or construct, new lines between known material. And after all, this is what all production of knowledge is about. The filter metaphor is meant to illustrate the borderline between scientific knowledge and political action. The NGOs, perhaps with the exception for Bellona, have their strength in transforming scientific facts into political issues.

It is obvious that the NGOs play an important role in building and presenting competence and knowledge, but their function seems even more important when it comes to "translating" knowledge into practical activity. In this context, the organizations function as a mediator between knowledge and practice. This seems to be the main task for all the NGOs, whether the recipient of their information is the general public, the politicians, the civil servants or the industry.

The NGOs' increasing role in knowledge production is not restricted to the environmental field, but relates to a more general process which has come to be called "the new production of knowledge" (cf Gibbons et al 1994). According to this view, the transformation in knowledge production in recent years has led to a blurring of the borders between science and society both in an organizational, economic and political sense, and there are thus opportunities for non-governmental organizations, as consultants (like Bellona) to take more direct part in scientific and technological innovation.

Another central role for the NGOs is as distributors of information. Most NGOs produce newsletters or magazines to their subscribers. Of course some of their readers are more important than others; both the FIVH and Bellona has underlined the importance of reaching the politicians. E.g. Bellona distribute "Bellona Magasin" to all the members of the Parliament. Among members of environmental organizations, such newsletters are regarded the main source of information on environmental issues (even more important than radio, television or newspapers). In this regard, combined with the importance of the organizations' own knowledge production, we might say that the NGOs represent a different channel, or vehicle, for disseminating environmental insights to the general public.

References

Bellona (1996) *Lepse-prosjektet*, Bellona facts sheet no. 14.
Berntsen, B. (1994) *Grønne Linjer*. Natur- og miljøvernets historie i Norge. Grøndahl Dreyer.
Brand, K.-W. (1995) 'Entering The Stage: Strategies of Environmental Communication In Germany.' In *Framing And Communicating Environmental Issues*, Research Report, Commission of The European Communities, DGXII, edited by K. Eder. European University Institute/ Münchner Projektgruppe für Sozialforschung.
Dahl, T. (1996) *Ordering Nature. Environmentalism as a Cultural Phenomenon*, STS report no. 30.
Diani, M. (1995) *Green Networks*. Edinburgh University Press.

Diani, M. (1997) Organizational Change and Communication Styles in Western European Environmental Organizations. Paper presented at the ECPR joint sessions of workshops.

Donati, P. R. (1996) 'Building A Unified Movement: Resource Mobilization, Media Work, And Organizational Transformation In The Italian Environmentalist Movement'. Pp.125-57 in *Research In Social Movements, Conflict And Change*, vol.19, edited by L. Kriesberg. Conn.: JAI Press.

Endal, D. (1994) 'Citizen Mobilization for Environmental Protection and Sustainable Consumption'. In *Steps towards Sustainable Consumption. A presentation of selected Norwegian initiatives*, Alternativ Framtid; report 2/94.

Gerholm, T.R. (1996) *Brev till Det Naturliga Steget*. Timbro.

Gibbons, M., et al (1994) *The New Production of Knowledge*. Sage Publications.

Gustafsson, M. (1993) *From Biocides to Sustainability. Swedish Environmentalism 1962-1992*. Centre for housing and urban research.

Jamison, A. (1996) 'The Shaping of the Global Environmental Agenda: The Role of Non-Governmental Organizations'. In Scott Lash, et al. eds.: *Risk Environment Modernity*. Sage Publications.

Jamison, A., R. Eyerman and J. Cramer, with J. Læssoe (1990) *The Making of the New Environmental Consciousness: a comparative study of environmental movements in Sweden, Denmark and the Netherlands*. Edinburgh University Press.

Jordan, G. and W. Maloney (1997) *The Protest Business*. Manchester University Press.

Kraft, S. (1997) *Rörelsens gröna ansikten*. Timbro.

Lem, S. (1996) 'Framtiden i våre hender: Punktmiljøvern eller samfunns forandring?' in Strømsnes and Selle (eds).

Lundberg, F. (1996) *Går miljömålen att förena? En skrift på uppdrag av fem svenska miljöorganizationer*. Fältbiologerna.

Lundqvist, L. J. (1996) Sweden in *Governing the Environment: Politics, Policy, and Organisation in the Nordic Countries*, edited by Peter Munk Christiansen. Nordic Council of Ministers, 1996:5.

Mullally, G. (1995) 'Entering The Stage: Strategies of Environmental Communication In Ireland'. In *Framing And Communicating Environmental Issues*, Research Report, Commission of The European Communities, DGXII, edited by K. Eder. European University Institute/ Centre for European Social Research.

Nilsen, K.E. (1996) 'Miljøstiftelsen Bellona: Døgnflua som overlevde', in *Strømsnes and Selle*.

Nilsson, B. (1996) Miljörörelsen, Volvo och bilismen. En jämförande studie med inriktning på social och teknisk förändring av hur tre svenska miljöorganisationer vill lösa bilismens miljöproblem och hur biltillverkaren Volvo vill lösa dem. Humanekologiska rapporter nr. 24. Avdelningen för Humanekologi.

OECD (1996) *Environmental Performance Review Sweden*. Organisation for economic cooperation and development.

Petersson, O. and D. Söderlind (1993) *Förvaltningspolitik*. Publica.

Robert, K.H. (1994) *Den Naturliga Utmaningen*. Ekerlids Förlag.

Rothstein, B. (1992) *Den korporativa staten. Intresseorganisationer och statsförvaltning i svensk politik*. Norstedts.

Selle, P. (1996) *Frivillige organisasjonar i nye omgjevnader*, Alma Mater.

Selle, P. and K. Strømsnes (1996) *Environment, Organization, Democracy: The Case of Norway*, paper prepared for the ISTR 2nd International Conference in Mexico City, July 18-21 1996.

Sevareid, H. (1996) 'Greenpeace Norge: En organisasjon uten fotfeste?' in Strømsnes and Selle.

Sjöberg, F. (1996) 'Yogin och konsulten – om den svenska miljörörelsens vankelmod'. *Framtider 4/96*. Institutet för Framtidsstudier.

Strømsnes, K. and P. Selle, eds. (1996) *Miljøvernpolitikk og miljøvernorganisering mot Dr 2000*, Tano-Aschehoug.

Sæten, A.J. (1991) *Miljöstiftelsen Bellona. I spenningsfeltet mellom statlig miljøforvaltning og frivillig miljøbevegelse*, University of Bergen; graduate thesis.

Szerszynski, B. (1995) 'Entering The Stage: Strategies of Environmental Communication in The UK'. In *Framing And Communicating Environmental Issues*, Research Report, Commission of The European Communities, DGXII, edited by K. Eder. European University Institute/ CSEC, University of Lancaster.

Tarrow. S. (1994) *Power in Movement: Social Movements, Collective Action and Mass Politics*. Cambridge University Press.

Uhrwing, M. (1996) *Att medverka eller icke medverka. De svenska miljöorganisationernas påverkansstrategier gentemot staten*. Statsvetenskapliga insitutionen.

Öberg, P.O. (1994) *Särintresse och allmänintresse: Korporatismens ansikten*. Doktorsavhandling Uppsala universitetet. Almqvist & Wiksell.

Økern, B. (1993) *Makt uten ansvar*. Grøndahl og Freyers forlag.

Chapter Four

Participation by Mandate: Reflections on Local Agenda 21

By Jose Andringa, Marco Giuliani, Patrick van Zwanenberg, and Magnus Ring

1. Finding the Public in Local Agenda 21 Activities

Diverse actors increasingly see public participation as a crucial component of sustainable development initiatives. In large part, this reflects the influence of environmental organizations at the Rio Earth Summit in 1992, and in particular, the so-called Agenda 21 which was formulated at the Rio conference, with its explicit commitment to greater public inclusion in policy making. At the same time, calls for greater consultation with the public have also dovetailed with national deregulatory trends and neo-liberal policies.

These broader developments, involving a rolling back of state direction and influence, have underpinned an increased emphasis on the benefits of partnerships between the public and private and voluntary sectors. Some of the more energetic attempts to involve the public and "stakeholder" groups in policy-making have been taken up by local government under Local Agenda 21 (LA21). Indeed LA21 has rapidly evolved into an umbrella term for a wide range of initiatives organized by local governments throughout Europe in which principles, targets and policy options for local sustainability have been developed.

There appear to exist at least two very different rationales for encouraging greater public engagement in policy-making in the post-Rio world. The first of these starts from the premise that many of the changes that are assumed to be integral to moves to greater sustainability require changes in public behavior as well as government policy. Here public participation is viewed primarily as a procedural good, a means by which the wider objectives of sustainability can be operationalised. This view is typically assumed by national government. For example, the UK launched a campaign called Going for Green in 1995 which targeted individual households, seeking to engender lifestyle changes largely through information provision. It assumes a deficit in public knowledge and understanding of environmental issues which, once filled, will result in changed behavior on the part of the public.

The second rationale tends to see participation as more of a substantive achievement in its own right. It rests on a more radical conception of what sustainable development entails and is more prominently held amongst some environmental NGOs rather than government or business. Initiatives with more deliberative, bottom up forms of participation are, however, in tension with many of the assumptions embedded within dominant approaches to dealing with the environment and sustainable development. Such approaches typically assume, for example, that definitions of what objectives and goals are or are not sustainable can be reliably determined by scientists and other experts and then implemented in conjunction with wider publics. But such an approach conceives of the public in instrumental terms, refusing to acknowledge that what does or does not count as sustainability is a negotiated process. The cleavage between an instrumental and a substantive commitment to participation – between, as it were, the public as consumers versus the public as citizens – has very different implications for how, in practice, initiatives such as the Local Agenda 21 process actively construct and involve the public in decision-making processes.

Although LA21 has only had a relatively short life span thus far, a range of questions seems pertinent to any evaluation of this attempt at stimulating public engagement in sustainable development. Firstly, and most straightforwardly, one can ask questions about the scale and nature of the various activities promoted by local government. For example, what sort of projects has been established and how widely have they been taken up? Secondly, one can consider the impact of LA21 initiatives. Do projects result, or appear to be resulting, in a meaningful reallocation of resources or are they more symbolic in effect? If the former, what sorts of actors, policy options and technological changes are being influenced by this process? If the latter, what implications might arise? Thirdly, and perhaps most pertinently, who is becoming involved in the various LA21 initiatives, how are they involved, and why are they involved in LA21?

This last question has three separate components to it. The first part refers to unpacking the public. Are participants, for example, serving as "ordinary" members of the public or are they representatives of community organizations and environmental NGOs who may have been involved in policy-making to some degree prior to LA21? To what extent do participants reflect the existing local population, along lines of class or ethnicity, for example? The second part of the question refers to how, precisely, the public are being constructed through the various institutions and programs that are being developed within LA21.

The third part of the question refers to how we might understand public responses to those initiatives. If, for example, the public or elements of the public are unenthusiastic about LA21 and sustainability, why might that be so?

In this chapter, we will briefly describe some LA21 activities in Britain, Italy, Sweden and the Netherlands in order to show how the attempts to resolve these tensions have varied significantly from country to country. LA21, we might say, has served different purposes in different countries, depending both on the level of environmental awareness, as well as on the needs and conditions of local governments. In order to exemplify and give some life to our general "national" observations, we have explored specific cases of local Agenda 21 activity in Sweden and the Netherlands, based on interviews with local participants. Our reflections on Agenda 21 are thus based both on literature reviews, as well as a certain amount of fieldwork and empirical investigation. Our aim has been not to disparage LA21 activity, but rather to identify some of its inherent qualities and limitations.

2. LA21 in the UK

Even though LA21 is a non-statutory obligation it has been pursued with vigor by UK local government. Over 70% of local authorities are estimated to be pursuing a LA21 process (LGMB 1997), with activities ranging from green audits, the adoption of environmental management systems, the development of sustainability indicators to new forms of – sometimes quite novel – decision-making procedures. One reason for the considerable interest by local authorities is probably a reaction to the undermining of local government powers over the last 15 years (for example, through the privatization of local government services, cuts in central government grants and the centralization of decision-making authority). LA21 has thus provided an arena in which local government could again play a distinct and meaningful policy role (Young 1997).

LA21 initiatives may also be seen as a local government response to a more general sense of alienation from political institutions. Indeed elected local authority members have recognized the potential for LA21 to revitalize local democracy and help articulate a new mandate for local government. (Tuxworth 1996).

Many LA21 activities in Britain have been concerned with integrating sustainability principles into other policy areas such as waste management, transport strategies and, somewhat less so, in sectors such as housing, education and investment strategy. Local government has also embarked on programs of awareness raising using existing communication techniques. For the most part, traditional instruments for incorporating the public's views into local government sustainable development strategies have been relied upon such as public consultation, questionnaires and public meetings.

Yet, a significant interest has been taken, at least by some local authorities, in broadening democratic participation and community involvement in these processes (Young 1997). About 50 to 60 of the 478 local councils in the UK have aimed at a more bottom up strategy in which local communities are actively involved in developing agendas for sustainability rather than the more conventional top down strategies of imparting information and asking for input into a pre-framed policy agenda. These more novel deliberative procedures include, for example, the development of surveys by local residents, arts-based approaches, visioning techniques, the use of round tables and "planning for real" exercises (Tuxworth 1996). Many of these bottom up approaches have, however, proved more rhetorical than real, since, when it comes to practical decision-making, councils appear to be reluctant to change their agendas and styles of work. Furthermore, there is a tension associated with these more deliberative processes in which some degree of decision-making power is devolved to local communities. Underpinning such LA21 initiatives is an implicit suggestion that traditional forms of representation (i.e. elected councilors) are not adequate to reflect local interests.

It is difficult to get a reliable picture of which stakeholder groups are actually involved in the LA 21 process. The Local Government Management Board's 1997 review of LA21 activities does not provide that information, although it does conclude that "sometimes it has been difficult to engage the real community beyond the pressure groups" (LGMB, p. 74). Even though surveys of public opinion suggest a consistently high level of concern about the environment across both class and age, the same surveys also suggest that public commitment to making lifestyle changes in favor of the environment has remained at relatively low and constant levels since the late 1980s (Worcester 1997). Indeed, the concept of sustainable development appears to excite little interest beyond environmentalists; most of the public has never heard of the term (Macnaghten and Jacobs 1997).

Some pointers to why many lay people have not been particularly enthusiastic about sustainable development, and thus perhaps why initiatives such as LA21 may have found it difficult to engage with ordinary members of the public, can be found in recent qualitative research on how people feel about environmental issues. Harrison et al (1996) suggest that the reasons why residents of Nottingham were inhibited from accepting wider environmental responsibilities included skepticism about the validity of environmental rhetoric and expert claims, especially in relation to personal and local knowledge, and the lack of trust relations between lay publics and government, the latter of which were not perceived to have any genuine commitment to tackling environmental and social problems.

Similar findings emerged from a study on public responses to proposed "sustainability indicators" in Lancashire that were being piloted in connection with LA21. (Macnaghten et al 1995). Such indicators (covering a wide range of environment, economy and quality of life areas, for example, with indices such as air quality, levels of recycling, acres of woodland, crime levels, employment and so on) are designed as a managerial tool that allows local government to monitor performance in service delivery. They are also intended to play a role in political objective-setting insofar as they can assist in foregrounding environmental questions in decision-making processes. Finally, and perhaps most significantly, they are also intended to promote public communication and participation. The study suggested that people were extremely skeptical as to whether central and local government or business could be trusted to promote sustainability; that environmental information emanating from those sources was similarly distrusted; and that a lack of agency in relation to that information affected the extent to which people felt it worthwhile to act upon that information. The study suggested that indicators were unlikely to command public confidence unless they reflected local people's own knowledge and were meaningful at a local level. Indeed, many people's concerns did not readily lend themselves to measurement. Thus indicators were most likely to be effective if they were developed in consultation and negotiation with the public rather than as a scientific top down procedure in which publics are treated only as consumers of environmental and social information.

In both studies the efficacy of collective action and indeed the democratic basis of UK society were thrown sharply into question by the research participants. Such wider considerations may well entail a lack of demand for participation in LA21 activities. Indeed they pose a considerable challenge for local governments who arguably need to rebuild trust and public identification with local government and devise new institutional and procedural means for genuine open-ended public participation in negotiating the definitions, goals and strategies for achieving sustainable development.

It is too early to judge the effects of LA21 activities on the shape of councils' LA21 plans (most of which are not yet completed) or indeed on strategic local government plans and the shape of council budgets (Young 1997). It is also difficult to obtain information on how LA21 activities influence developments outside of local government. The Local Government Management Board's 1997 review of LA21 activities concludes that "there is a widespread difficulty of involving local businesses in the Local agenda 21 process." (LGMB, p. 74) If that is indeed the case, LA21 may be building up a series of expectations that local government on its own is not able to meet. Measures such as shifts towards more environmentally benign energy supplies and improved public transport require capital investment of a type that local government is not in a position to provide. Moreover, without central government commitment in areas such as transport strategy (and indeed with most of the transport industry in private hands) local government, on its own, lacks sufficient powers to reallocate significant levels of resources, certainly of the kind necessary to realize sustainable development.

3. Local Agenda 21 in Italy

In this section, we will briefly review some common features of a few experiences in sustainable planning in Italy at the municipal level. More than extensive and in-depth case-study reconstruction, our "inspection" will discuss more or less successful experiences of Local Agenda 21.

At the end of 1996, there were between twenty and thirty Italian local authorities which had signed the Aalborg Charter (on sustainable cities). This quantitative value is not so different from that of other European countries (with the exception of the United Kingdom), but, also not unlike those other countries, it hides a very disparate reality. Many municipalities have signed the Charter only as a symbolic commitment towards the environment, and only a few of them have tried to implement its goals. Though the campaign for sustainable cities implies the consensual construction of long term action plans, and it is probably premature to evaluate thoroughly the adopted programs and the achieved progress, the fact that the national coordinator of the Local Agenda 21 is aware of only a few real cases of sustainable local planning justifies some skepticism on the overall success of the operation.

Nevertheless, some cities have progressed on the path first defined in Rio. The municipality of Bologna has received an "Ecolabel" at the Lisbon conference (a follow-up to Aalborg), and other local authorities like Modena, Venezia, Milano, but even Torino, Roma, and Genova are acting much in the same spirit. If the ultimate effects of these policies will be measured only in the next millennium, we can already look at the processes in progress, especially in terms of innovation and public participation. Following the logic of the "Most-Similar-Systems", we can try to distill from these policy histories some of the elements which have supported the few (more or less) positive Local Agendas 21, while controlling their peculiar vicissitudes.

Scanning through the Italian comparatively most advanced experiences in Local Agenda 21, the first and most obvious thing is that they are all characterized by a consistent cooperation between actors defined by different types of legitimization: political empowerment, administrative responsibility, technical expertise, representative accountability, etc. The Committee for Agenda 21 in Milano, for example, has been activated by a local administrator (who is also an environmental expert), and is coordinated by a person from the main Italian industrial association (and edits its journal dedicated to the environment) together with the local leaders of the two major national environmental associations (*Legambiente* and WWF) and the representative of the urban district committees. Bologna, which has a long tradition of interaction between local administrators and environmentalists, has for its Agenda 21 activity, hired a consultant research institute which evolved from that experience (*Istituto di Ricerche Ambiente Italia*). Venezia followed a similar path: on the joint initiative of the municipality and the province, and the consultation of local associations, it employed the external expertise of the *Fondazione Eni Enrico Mattei*. Without this cross-linking, the process of "sustainable planning" did not take place or was broken in its infancy. This certainly happened for the municipalities which responded only symbolically to the Aalborg Charter, but even cities like Genova, after a promising start, seem to have suffered from a lack of networking. From this, we can draw a conclusion that sustainable processes cannot be governed or managed bureaucratically as an attempt to reduce their complexity, but have to exploit complexity through the contribution of different kinds of competence.

A few conditions seem to have favored the effective creation of local connection among actors belonging to different domains: the possibility of relying upon already established (often personal) relationships, well-organized and scientific-oriented environmental associations, a cooperative bureaucracy, some albeit differently motivated political commitment, the appointment of external experts at an early stage of the process, the use of different communication channels to the citizenry (but not starting from a *tabula rasa*), and the presence of a "policy entrepreneur" activating the process, supporting it through its unavoidable crises and misfortunes and translating the multiform languages spoken by disparate actors into a common discourse.

Though the scale of environmental problems which have to be tackled by small cities is certainly different from that of a big metropolis, nonetheless the transversal character of sustainable planning requires even at that level a consistent cooperation from different policy actors, and especially of environmental associations. Unfortunately, environmentalists often are not able to continuously participate in complex policy-making processes. The risk is that environmentalists revert to more oppositional practices, without taking advantage of the possibility to have an enduring influence upon the whole local policy-making activity. Except for cases like Milano and Roma, where green movement organizations are sufficiently "equipped" to intervene directly, in other cities, like Modena and Bologna, they had to rely upon the contribution of friendly experts – green think tanks or academics.

Whereas bearing professional competence may increase the effectiveness of environmentalists' direct influence upon the process, it may also have unexpected shortcomings, like that of messing up the roles and of increasing the opposition, or antagonism, from the local bureaucracy. This happened for instance with the Local Agenda 21 of Modena, where the process had to be, in a certain sense, re-started in order to gain the cooperation of a suspicious local administration. The initial employment of an external research institute known to be "green", fostered an initial opposition from the municipal bureaucracy, which slowly started again to cooperate only after another consultant had been hired to set up a "more local" agenda. This may depend on the fact that "insiders" are less used to the argumentative style used in these networks, and, not being able to rely upon more formal rules, are more skeptical towards the scientific contribution possibly made by these external institutes.

More generally, a cooperative bureaucracy seems to be crucial in order to pass from the preliminary investigative stage to a more operative one. This is probably one of the differentiating elements between the experience of Bologna, which could rely upon a tradition of consensus building, and that of other cities which have not entirely fulfilled the initial expectations. Milano is probably the clearest example in this regard, with a process interestingly guided by environmentalists and economic interests together, but which has had nothing to do with the local administrative "machine".

The origin of the Agenda 21 of Milano was in fact mainly stimulated by what can be considered a policy entrepreneur: a personal effort by an independent member of the town-council shortly before an election, together with his attempt to initiate a wide-ranging project as a well-known expert in environmental matters. From a certain point of view, the coincidence of political and policy aims favored an innovative approach to the problem, letting it be governed through a concerted effort from the most interested "stakeholders" (environmentalists, entrepreneurs, town-districts): the "externalization" of the process permitted greater degrees of freedom but, at the same time, made it completely alien to the local bureaucracy. This entrepreneurial character favored the links between economic and civic domains, but completely bypassed the bureaucratic. For this same reason, the "policy entrepreneur's" departure from the council after the election led to the demise of the whole venture: a good start, but with only marginal results.

Some kind of "entrepreneuriality" is a common element of all the policy histories we are here reviewing: besides Milano, even the cases of Genova, Modena, Roma and Bologna present this aspect, but the overall effect depends more on how this activism is routinized than from how it is displayed. It is the innovative character of the concept of sustainability in itself which fosters this kind of changed behavior, but problems normally arise when confronted with the necessity of maintaining a complex network of very different actors working together.

Finally, since the goal itself of Agenda 21 is to produce a consensual long-term environmental plan via the inclusion of the public interest, we must pay attention to how the institutions formally involved in the project have maintained the process open to the citizenship participation. In this regard, local experiences diverge both in the substance as well as in the means used. As far as the content is concerned, whereas no municipality has simply accepted suggestions at every level, the timing of the encounters with the public have been quite different: those who have opened the process quite early, e.g. before the preparation of a first set of eco-indicators, like Modena, have been often delayed even on minor aspects; a pre-existent frame, still open to variations, seemed to work better, as in the case of Bologna and, more recently, with Venezia. Both these municipalities, in the latter case with the help of local associations, have even experimented with new means to keep the public aware of the project and of the open initiatives: in addition to more traditional encounters, *fora*, conferences, laboratories, committees, etc. they have established newsletters, web pages on Internet and "Agenda 21 awards" for micro-initiatives.

Local agendas are thought to be a key to achieving sustainable development for the 21st century. Most of the municipalities in Italy have not taken on this challenge, and many have done it only symbolically, simply subscribing, for instance, to the Aalborg charter. The fragments here reported all belong to more advanced policy histories, though their actual outcome is yet to be evaluated. Bologna has been internationally acknowledged for its activism, and, comparatively, is probably the most evolved Italian agenda, but even Venezia, which started slower, is proposing a project which is interesting both in the contents and in the method.

4. Local Agenda 21 activities in three Swedish municipalities

In this section, we will focus on particular local experiences with LA21, by considering how three Swedish municipalities have organized and carried out their LA21 activities. The three Swedish municipalities whose work with Agenda 21 we have studied closely are Växjö, Malmö and Lund, three municipalities that differ in many ways. Lund is an old university town, while Malmö is a large industrial city. Växjö, on the other hand, is the home to many small and medium sized companies, and now to an expanding new university, as well.

The three municipalities have each had different ways of organizing their work with Agenda 21. In Lund there is a so-called environmental delegation within the municipality, in which the Agenda 21 coordinators work. In Växjö, as well, the coordinator is directly under the municipal board, but is placed within the planning department, and in Malmö, the two coordinators are part of the environmental administration.

We have tried to develop a picture of how Agenda 21 activities are organized within the municipalities, and also what form the activities have taken. In addition, we have explored the relation to the national authorities and environmental organizations, which is why we have also contacted the central agencies of the National Environmental Protection Agency and the Swedish Society of Nature Conservation (SNF).

Malmö

The environment administration in Malmö has established an Agenda 21 office where a proposal for a local Agenda 21 document has been developed. The aim is to create unity concerning what is called "basic principles, comprehensive goals and visions for the endeavors to create a long term sustainable development." This document is then sent for consideration to organizations, companies, administrations and individuals. The ambition is that all the inhabitants of Malmö should have the opportunity to take part in the document and be able to express their opinions before the final document is composed. Thus, the process is time consuming. The hope is that once there is unity concerning the cases, the work can be initiated. Furthermore, the city of Malmö has started a process of educating environmental supervisors and environmental spokespeople among its employees (something that is common in other municipalities as well). The strategy for the municipality's Agenda 21 coordinators is often to provide training for some of their "own" first, the idea being that they will later circulate their knowledge in the context of their daily work. This is a common strategy among the coordinators with whom we have spoken, and various educational efforts are often seen as the best kind of action.

> "Yes, the individual action that works the best, always, is to give a seminar, some training, where someone from the outside comes in to speak about an interesting topic. Then there will be a step forward each time. So that sometimes I think I should do nothing but organize seminars."

As for the activities of the environmental organizations in Malmö regarding Agenda 21, the project "the Green House", initiated by several of the local environmental NGOs, stands out. The main idea of the project is to create an information center and gathering place for environmental issues. In addition to the municipal government activity with Agenda 21, there are also other groups in Malmö which are working toward the same goal, and with the same issues. One of the most vocal has been Kirseberg's local Agenda 21 group, which works with environment issues of concern for that particular district. There are also other local groups working to improve the neighborhood environment and create environment centers, such as the Möllevång group and Theater X, both of which are active in areas sometimes described as "problem areas". In these cases, the social aspect (lower crime, higher security, etc.) might be more important than the environmental issues and environmental goals. In comparison with Växjö, however, the Agenda 21 activities of business firms seem to be smaller in scale. But there are examples, nonetheless, of companies in Malmö which are training their employees in environmental issues, and developing plans for easing the pressure on the environment.

The city of Malmö's proposal for an Agenda 21 program/document was presented to the public at an open meeting at the beginning of January 1997. However, it was not the municipality that organized the meeting, but the left party's local section in Malmö. The Agenda 21 coordinators took part, and spoke of the importance of looking over, and adjust the municipalities activities, so that it could be a model. This is an ambition shared by all of the coordinators we have spoken to in our study. As in the case of training the employees, the assumption is that by being a good example and model, the rest of the society will also get involved in working with Agenda 21. Furthermore, it is hard in this case to see a correlation between example, influence, and results, since it is an indirect process, where there are no direct efforts to influence, for instance, companies. Malmö's local programs emphasize resource management and lifecycle consciousness. There is also an interest in renewable energy technology. At a public meeting to discuss the final document, the conflicts between the different actors in Malmö were quite apparent. For instance, the Kirseberg group, mentioned above, found the document too vague. In this lies the dilemma between, on the one hand, giving recommendations and setting an example, and, on the other, trying to influence more directly through for instance legislation (on a national level), fees, or other concrete actions.

Lund

Lund has established an environmental delegation where the two Agenda 21 coordinators work. The environmental delegation in Lund is placed right under the municipal board, which has appointed the seven politicians who make up the delegation. In this case, the Agenda 21 coordinators are the employees of the delegation. But the politicians in the delegation are not all members of the municipal board, which has caused the municipality some problems in terms of who is to make the decisions. Therefore, the main role of the delegation has become advisory. The environmental delegation has no formal power, but rather a coordinating function within the municipality. In Lund the environmental delegation has presented a proposal for a local plan of action for Agenda 21 which was approved at the end of 1997.

The strategy in Lund has largely been to conduct the work with Agenda 21 at the district level, which leads to further decentralization. Each district now has one responsible person on Agenda 21, and a number of environmental representatives have also been trained at the various work places in Lund. The responsible person gathers information from the various environmental representatives within the city district, and through these representatives, a kind of network is being built. This strategy is considered more effective than trying to construct something more centralized within the municipality.

The overall picture emerging is that environmental action still depends on popularity: i.e., that there is a public interest in being environmental. But the coordinators' impression is that there is still a long way to go, and that it becomes all the more difficult when "uncomfortable" decisions must be made. It is seen as a process whereby the citizens must both be educated and engaged in long term environmental activities. In Lund, as in Malmö, the focus has been on educating so called environmental advisors and environmental representatives at different levels within the municipal organization.

An organizational solution has been reached by way of much decentralization, and the focus has been on providing the local population and the municipal organization with information. Within the city there are also a number of other related actors. The most important of these are the Rio-group at Östra Torn, an activist-group created around a neighborhood as a direct consequence of Agenda 21 and the Rio-conference of 1992. Within the municipality, there are also various neighborhoods working with environmental issues, as well as the Environment Library, which is a non-profit project compiling literature concerning the environment. The local section of SNF is the established organization with which there has been most contact in connection with Agenda 21. On the other hand, collaboration with SNF in Lund in general is described as being locally oriented. Furthermore, in Lund the national committee has been found absent. But we have also seen how the work in Lund on the part of the municipality has been geared toward further decentralization, and how the main work consists of spreading information, and in the long run, engaging the citizens of Lund and in this way trying to make their behavior "environmentally adapted". The main idea behind this is that a certain degree of acceptance must be established among the citizens before behavior can be fundamentally changed. The impression so far is that the issue has been to initiate a process in order to establish environmentally attitudes or values on a "micro-level". Here it is clear how the principle of participation on the part of the citizens has been taken seriously, but also how a comprehensive national perspective/contact network has been lacking.

Växjö

SNF has selected Växjö to be an "environmental municipality". What this means in practice is that SNF has a highly developed collaboration with the municipality. The collaboration primarily takes place between SNF's central office in Stockholm and the municipality's Agenda 21 coordinators.

Växjö is especially interesting in that it engages in a kind of alternative program of environmental policy where there is an attempt to involve more people than those from the already established groups and authorities. Above all, the outreach to companies and the endeavor to influence them through networks has been actively pursued. But the companies on their part have also influenced the municipality through its environmental office. The activity which has gotten most attention is the so called "environmental coffee", where business-people of the city meet and discuss environmental concerns over a cup of coffee on a regular basis with or without representatives from the local authorities. "Environmental coffee" has been taking place for two years and has led to established contacts with companies as well as with various official institutions. "Environmental coffee" can be seen as a network created to carry out environmental policy. The coffee is an attempt on the part of the city's environmental authorities to increase the dialogue between companies and authorities (interestingly enough it came to existence as a response to an initiative taken by those responsible for environment issues at two companies in Växjö). On the part of the municipality, "environmental coffee" is in its extension an attempt to reach out to as many people as possible in their everyday context and thereby influence them to show concern for the environment. But also more concrete projects have been borne out of the creation of networks resulting from "environmental coffee", such as using and developing alternative fuel at one of the large transport firms based in Växjö. From the perspective of the municipal authorities, the network constructed as a result of "environmental coffee", has been successful, consisting of some two hundred people and over 100 companies.

Crucial in terms of research and progress concerning environmental issues, is the local energy company, VEAB (Växjö Energi AB) and its shift from fossil fuel to bio-fuel in the production of electricity and heat. VEAB is considered a forerunner in the transition to renewable energy sources (Löfstedt 1995). The company actively supports research and development in the area of bio-energy, and as a part of this, the company has contributed to the founding of a center of bio-energy at the University of Växjö. Here various links are established between the municipality, companies and research projects. There is also a collaboration with TPS (Technical Processes AB in Studsvik) concerning the process development within combustion technology and the classification of bio-fuel. The Center of Bio-energy at the local university is a result of an initiative from several companies and institutions, such as the municipality, the county administrative board, state authorities, ABB Fläkt, and the regional forest-owners. In this process, the municipality and various companies have been the main driving forces. VEAB has worked with environmental adaptation for quite some time, and the efforts have been focused on utilizing the sources of bio-fuels available in the area. Just as it has been only natural to develop wind power as an alternative energy source in Denmark, it has been natural to utilize the (renewable) forest as an energy resource in forest communities such as Växjö.

Växjö's Agenda 21 coordinators collaborate to a large extent with the central office of SNF. Both parties experience the collaboration as fruitful and this is the primary example we have found of how an environmental organization and local "bureaucrats" have been able to come up with a common strategy for carrying out a successful Agenda 21 enterprise. As one of the SNF representatives puts it, "...if we can make the local politicians run our errands, this is the unofficial strategy, make them say the same things, then it is much, much louder in the ears of the national politicians." There are a number of Agenda 21 related activities in Växjö, such as projects in various neighborhoods, local interest organizations etc., which are not affiliated with this overarching strategy developed in the relation between SNF and the Agenda 21 coordinators in Växjö.

The municipalities that we have studied show both similarities and differences in terms of their organization of Agenda 21 activities. Lund has embraced and pushed the decentralization process. Energy has been invested in informing and educating the municipality's citizens and staff, in an effort to ground Agenda 21 at a grass-roots level. A pronounced strategy is to prepare a more radical environmental policy for the future. Educating and informing, as well as creating a decentralized organization/network within the municipality are measures believed to lead in this direction.

Malmö has developed a similar strategy. The difference is that in Malmö, the city's organizational structure has not been changed, as in Lund, by creating a special environmental delegation. Instead, the responsibility for Agenda 21 has been placed on the municipality's environmental management office where the two coordinators are positioned. This has both advantages and draw-backs. Pre-existing knowledge and networks can be extended, but the risk of Agenda 21 disappearing into, and being influenced too much by, current bureaucratic structures (which might slow down the process) is obvious.

Växjö has to a large extent focused on "adjusting" the municipality's activities to long-term goals for sustainable development. This is a consequence of the collaboration with SNF, and also of the local political consensus regarding environmental issues. Information and education are crucial here too, while bigger and more long-term decisions have also been made, most importantly that of becoming a fossil fuel-free municipality. Efforts to co-operate with local companies, both in terms of long-term environmental decisions and in a broader sense, have also been successful.

National coordination
In Sweden, the work with Agenda 21 has for the most part been assigned to the municipalities and various environmentally oriented (often local) organizations, but this does not mean that there is no activity at all on the national level. The Environmental Protection Agency has a key position as evaluative authority, and also reviews project applications and distributes a large part of the grants for various Agenda 21-related projects (some seven million Swedish crowns per year over a three year period, 1995-1998). Furthermore, in 1995 a national Agenda 21 committee was created, consisting of representatives of the political parties that are represented in the Parliament, as well as of representatives of various societal sectors.

Most of those involved seem to agree that the work with Agenda 21 has developed well in Swedish municipalities. But the opinion is also expressed that the municipalities are "ahead" of the government and that the national authorities should relate themselves more closely to the actual development taking place in the municipalities. Some fifty of Sweden's two hundred and eighty eight municipalities have come so far in their work that they experience what SNF calls "National Obstacles".

In its report, *Nationella hinder för Lokala Agenda 21* (National Obstacles to Local Agenda 21), SNF and the youth network organization q2000 point out that a national Agenda 21 is by no means to be seen as the sum of all the local efforts. They call for a stronger effort on the part of the government, the Environmental Protection Agency, and the national committee. These national efforts must "do something more than gather good local examples," as it is said in the report. The organizations consider a more forceful national policy to be a precondition for success at the local level, as well as in regard to a national impact for Agenda 21. The report, as well as the results of our interviews, indicate that a gap has opened between the national and the local, partly as a consequence of the government's decision to give the municipalities the overall responsibility for Agenda 21.

Since Agenda 21 activities are meant to be organized by the municipalities and their citizens, they are largely oriented to specific local projects of great diversity. In addition, municipalities have placed the work with Agenda 21 within different municipal agencies or authorities. The size of the grants given to the work in different places also varies, Lund being the city whose government has proportionally given the most money to activities pertaining to Agenda 21. Thus, both the organization of the work and the resources devoted to it have varied greatly among different municipalities. The demands on national coordination have also increased over time. Some municipalities find that not only do they lack financial resources, but perhaps especially knowledge resources and access to the range of expertise that exists in "environmental Sweden".

It is in primarily two areas that national aspects appear to be important, and actual obstacles for the work pertaining to Agenda 21 have been identified: 1) legislation, where laws concerning public procurement and producer responsibility are seen as limited, and 2) exchange of information, which is seen as problematic. There is a widely felt need for information from the national level; many of those involved express the view that political decisions taken on the national level are often detrimental to fulfilling intentions at the local level.

As we have seen, the Swedish Society for Nature Conservation (SNF) appears to have taken on the role of coordinating many of the Agenda 21 activities. This is noteworthy, since the SNF is a non-governmental organization. But already at the outset, the authorities wanted to place the work with Agenda 21 on as local a level as possible, which is why, according to SNF and q2000, national coordination has been lacking. This view is also affirmed by our interviews. As one of our interviewees put it, "I think that most municipalities working with Agenda 21 have not even known that there is a National Agenda 21 Committee, and if they have, it has no practical impact on the work."

According to information from the authorities, all municipalities in Sweden have now initiated Agenda 21 activities. But SNF also demands that there should be a meaningful national Agenda 21. What is needed, according to SNF, is a plan of action for how Sweden is to become sustainable. For, of course, it is problematic for SNF that the national coordination does not work well in these situations. As a kind of response to this, the association is directing the project "Sustainable cities" since 1993. The primary purpose of this project is to serve the municipalities and the local groups within SNF. SNF's goal in these contexts is to increase the possibilities of influencing decisions in local environmental policy from organized environmental groups as well as from individuals.

We here see how SNF as an environmental organization crosses over into what can be termed a "bureaucratic policy domain", at the same time as the association's environmental policy has its basis in civil society, or the civic policy domain. Although SNF is no official authority, the collaboration with the municipalities and the Environmental Protection Agency is so close that the role of the environmental organization approaches that of an authority. Therefore, one might say that SNF helps fill the gap between, on the one hand, the local and the national, and on the other hand the public and the policy makers. SNF could also be seen as a "consultant" in these situations, albeit with the important difference that SNF is not paid for its services. Therefore, SNF's board is currently considering decreasing the organization's involvement in these activities, partly for economic reasons.

This central role for SNF reflects the fact that during the course of the 1980s and 1990s the environment movement has diminished in importance as a grass-roots movement in Sweden and other countries, and instead been drawn into programs of "green consumption" and sustainable development. Among environmental organizations, there have been processes both of institutionalization and professionalization, with the different organizations competing for members and influence on a political and economic marketplace. The result is that the space between the civil society and the bureaucracy, between the public and the policy makers is becoming an open playing field for various entrepreneurial agents, such as consultant firms, research institutes, environmental NGOs, etc. This field could perhaps also be the breeding ground for a new "social movement", which at least some of those involved in local Agenda 21 activities imagine themselves to be.

But Agenda 21 is not a social movement in any meaningful sense: there is no alternative collective identity being established, no explicit political protest or campaign, and, perhaps more specifically, no construction of an integrative "cognitive praxis" building on new kinds of alternative ideas or knowledge interests, as was the case with the environmental movements of the 1970s (see Eyerman and Jamison 1991). Rather, sustainable development is a concept and cluster of knowledges and competences that has been imposed on LA21 activity. It has been provided as a kind of external, overall "frame" for activities, but without the active constructive and creative involvement of the actors themselves.

The lack of cognitive praxis is interesting in regard to the process surrounding Agenda 21, since much of the imposed relation between the bureaucratic and the civic domains is based on an external need for knowledge and expertise from the municipal coordinators. Agenda 21 lacks a cognitive praxis since there are not (yet) opportunities for constructing knowledge from below; rather, Agenda 21 comes from above, from pre-established structures and external conditions and experiences. The cognitive praxis which might be said to take place is related to the knowledge possessed by the established environmental organizations and experts. The Agenda 21 process is dependent upon existing knowledge, which in turn can be seen as a reason for why the established interests have acquired such a key position in the process.

It has primarily been SNF that has bridged the gap between the civic and the bureaucratic domains. Although there is plenty of initiative on the local level, these have not channeled to the policy makers. The municipal authorities, too, seem to have problems creating and sustaining contacts on a national level concerning the work with Agenda 21, with the exception of Växjö, which has established that relation through SNF. It is also remarkable that the social movements not specifically involved in environmental issues are absent in "Agenda Sweden". Thus, Agenda 21 is integrated into the Swedish corporatist political culture, and it is mostly in terms of pre-established roles that the civic and bureaucratic domains are communicating.

5. Local Agenda 21 in the Hague

After UNCED in 1992, the Agenda 21 received little attention in the Netherlands, and only a few initiatives were undertaken by NGOs and local authorities. That was the reason why the Platform for Sustainable Development (PDO) established a working group LA21 in 1993. This Platform (now called the NCDO) represents Dutch NGOs in the preparation of the annual report on Agenda 21 activities to the international Committee on Sustainable Development. The activities of the working group LA21 include the organization of LA21 meetings, so that municipalities and NGOs can share their experiences, publications about LA21, and forming advisory teams at the request of people who want to start LA21 activities. These activities show features of a national campaign to stimulate and support local initiatives. The NCDO strives for 100% of the Dutch municipalities to be working on LA21 by the year 2002.

From the very start, environmental groups, as well as the Ministry of Environmental Affairs, were involved in establishing the NCDO Working Group. Now, however, only non-commercial organizations with their own rank-and-file members are represented in the 15-member group, whereas formerly environmental advisory consultants were also involved. A recent initiative of the Group is to send to all municipalities a letter in which concrete textual suggestions are given for the election programs for the municipal elections in 1998. The letter advises municipalities and political parties to take up LA21 as an instrument for a sustainable future, cooperating with interest groups, companies and individual citizens. The letter also refers to a document with suggestions and background on LA 21, published by the NCDO. The NCDO has also established a fund in cooperation with the Ministry of Environmental Affairs, to support local groups' first initiatives towards a LA21.

In 1996, LA21 became a so-called task of choice by the Ministry of Environmental Affairs in the VOGM, or "Continuation of Support of Municipal Environmental Policy". The VOGM is a program that provides extra funds to help municipalities implement the targets of the National Environmental Policy Plan (NEPP) introduced in 1990. The idea was that after 1995 the funding would no longer be earmarked for environmental purposes. But an evaluation during the funding period raised doubts as to whether the environmental tasks were really well institutionalized in municipal organizations. The Environmental Ministry and the Association of Municipalities decided to introduce the VOGM to strengthen the role of municipalities in environmental policy. This new earmarked funding for 1996-1998 was to give municipalities better opportunities to make their own priorities. Municipalities could receive extra funding for four policy priorities out of a list of nine, of which LA21 was one.

About 140 municipalities chose LA21 as one of their four action points. However, the actual execution of this action point has not always taken place. Werner Sikken (of the NCDO) is critical of LA21 being one of the actions of this VOGM, since the character of LA21 is very different from the other action points. Because it is on the same list, it can seem to be an administrative task, like taking proper care of sewage. In the opinion of the NCDO, LA21 is far more than that. In addition, the VOGM is only focused on environmental policy, so the other dimensions of LA21, such as international cooperation, tend to be neglected in the implementation of this action point.

The Dutch Association of Municipalities (VNG) has not been so active as similar organizations in other countries. Traditionally the VNG supports its members in the execution of concrete (VOGM) tasks, and so they do not have much experience in the integrated and dialogue character of LA21. However, in 1996 they published a handbook on the subject, mainly written for civil servants, in which LA21 is explained and examples are given.

In the 1990s a crisis in local democracy and local administration in the Netherlands was recognized, and almost all Dutch municipalities have taken part in the so-called administrative renewal. The extremely low participation (61,5%) at the local elections in 1990 prompted these initiatives. It was realized that the public support and legitimacy for municipal authorities was in serious trouble. A study of voters' behavior indicates that citizens are not particularly interested in local politics, but take part in the local elections because they feel it as a citizen duty (Veldboer 1995). Local voters base their votes on national politics, instead of on the acts of local politicians.

However, it is not only a disinterest of citizens towards local politics that is involved in the "crisis" of local government. Citizens' expectations of local administrations have also changed. More and more citizens see themselves as customers and the municipality as a provider of high quality services. There is also a widespread feeling of being excluded from the formulation of policies and plans by the government. This has led to a number of efforts to try to "reinvent government" in the 1990s. Many initiatives have been undertaken and experiments carried out. "Administrative renewal" is a term that covers these initiatives and experiments directed to get more public support for policy and more participation of citizens in policy.

Local Agenda 21 initiatives fit under the umbrella of administrative renewal, in the specific policy arena of sustainability. This does not mean municipalities easily take up the concept of LA21. According to Sikken

(NCDO, interview 28 April 1997), local administrations do not have much experience in involving citizens in policy making. Alderman Van der Putten in the Hague has stressed the importance of public participation in the building of a LA21 "also seen from the viewpoint of administrative renewal: people should have the possibility to influence their immediate surroundings."

In May 1994, Van der Putten attended the European Conference on Sustainable Cities and Municipalities and signed the Aalborg Charter, thus committing The Hague to building a Local Agenda 21. As he put it later: "I attended this Conference because I was interested in the issue of sustainable cities. After being an alderman for one year at that time, I started to look at what my colleagues in other cities were working on, in order to learn from them" (interview, June 5, 1997). Van der Putten hardly knew anything about Local Agenda 21 at that time. He learned at the conference about this concept and also found out that achieving coherence in environmental policy was a challenge for The Hague. From the presentations of other municipalities Van der Putten got self-confidence: at home he was working on the same issues as these colleagues were. The Hague did not perform "that bad", but could compare itself rather favorably to these "sustainable cities".

After the Alderman came back from Aalborg, he established the Haagse Locale Agenda 21 (HLA 21). The first project leader and "spider in the web" was Bert Bovenkerk, later followed by Theo Kuijpers, both from the environmental department of the municipality. Officially, the project leader functions independently from the municipal authorities. He works directly under the responsibility of the Steering Board of the HLA 21 and coordinates between the Steering Board and the working groups. The only official role of the municipal authorities of The Hague was to initiate and facilitate the HLA 21 process. However, project leader Kuijpers has an office on the same floor as the environmental department and he is involved in the regular departmental communication structures. He thinks this is crucial to communicate the results of the working groups and to share experiences with his colleagues. The Steering Board is supposed to follow the process carefully and to advise when necessary. It also functions as a means to achieve broad communication and dissemination of results. Members of this Board are recruited from different policy domains: the chair J.M. Cramer (professor Environmental Management, University of Tilburg and senior researcher at TNO), S.J. van Driel, general director local government, E. Biesbroek (Regional director NV ENECO The Hague/Voorburg), J.C. Eikelboor (priest Petra Church/Johanneskapel), D. Bovekerk (market manager McDonald's), J. Schinkelshoek (chief editor *Haagse Courant* a regional newspaper), and S.M. Slabbers (landscape architect).

The mayor and four alderman are members of a so-called Sounding Board, and are kept informed on the progress of the working groups. The four alderman are from Spatial Planning (Noordanus), Traffic and Transport (Meyer), Public Health (Luyten) and Environment (Van der Putten). But according to Kuijpers, this Board has hardly functioned. However, Van der Putten and the other alderman do have at least an official role in the process. And we will see in the description of the temporal developments that especially Van der Putten is important in putting issues from HLA21 on the municipality's political agenda.

After alderman Van der Putten returned from Aalborg he had a discussion paper written on the content of the Aalborg Charter, which was discussed in the committee of the Municipal Council on Environmental Affairs. In January 1995 the Council agreed officially to implement the Aalborg Charter, specifically stressing the building of an LA21. Meanwhile the first preparation documents were already being written. The first document offered an overview of the latest municipal developments on the seven themes of the Working Groups and a first move for discussions in these Groups. A document inviting all citizens to take part in the LA21 process was published and sent to a large variety of groups and organizations in March 1995. In this document the organization structure and the themes of the working groups were also briefly described. A leaflet was distributed more broadly, via municipal offices and libraries.

In June 1995 the HLA 21 organized an open house, in order to attract members for the Working Groups and to get attention and publicity. The number of respondents who wanted to join the working groups turned out to be less than expected. Meanwhile the project leader searched for ways to involve more citizens and organizations. Eight working groups (80 participants in total) started in the fall of 1995. Although the project leader and his team tried to reach more people by sending letters to specific organizations and making phone calls, representatives of companies as well as lay citizens appeared hard to reach.

The activities of the Working Groups, or Theme Groups as they are also called, consisted of three main steps: an inventory of the state of the art of the given theme, an overview of possible solutions, and a list of possible actions. The ideas or projects are called *poekels*, the local term for "having a conversation or dialogue".

In early 1996, the projects began to take shape, and a first newsletter of HLA 21 was published. The groups were explicitly asked to check the contribution to sustainability and the expected public support of their ideas, and to search for possibilities to finance their projects. Because of the financial situation of The Hague, there was, however, no extra municipal money to finance the implementation of the *poekels*. An internal conference in March 1996 offered opportunities for mutual informing, and sharing experiences among the Working Groups and other participants within the HLA 21 activities. At the conference, some 60 *poekels*, varying from rough ideas to more worked-out projects, were presented. It became clear that some Working Groups only had a few (active) members and were struggling to find a meaningful way to work. The method of Traffic and Transport was seen as an example. At the conference it could be noticed that there was very little cooperation between the Working Groups.

After this conference, the Working Groups continued with renewed energy and the project leader and administration prepared a Public Day in June, when three thousand visitors attended in the City Hall. Alderman Van der Putten was given the "Poekelbook", and passed it directly to the alderman of Traffic and Transport and the alderman of Spatial Planning. He also opened a local Department of Greenwheels, a care-sharing initiative, thereby officially realizing one of the *poekels* of the Working Group on Traffic and Transport. Also a forum discussion was held between representatives of the municipality, the Body Shop, the Chamber of Commerce, the Triodos Bank, and the Ministry of Environmental Affairs.

Many of the Working Groups then stopped their activity, because they felt they had done their duty. Other Groups have continued in order to implement the *poekels* they proposed. A still active and promising Working Group is the one on Traffic and Transport, on which we will focus below. It should be noted, however, that the Working Group on Traffic and Transport has been exceptionally active.

The Working Group Traffic and Transport started with eleven participants stemming from, among others, the ENFB (the Real Dutch Cycler's Association), Working Group Church and Environment, a political party (D'66), the company Meeuwisse Nederland BV, and a few "ordinary" citizens. Meetings were held once a month, mostly in the *Haags Milieu Centrum*. In the first meeting, on the 10th of October 1995, only 3 people showed up: Ronald Bijl, Johan Apeldoorn and Peter Creemers. From the start, Peter Creemers was contact person and secretary of the Group, participating as an "ordinary citizen" while working as a volunteer at the ENFB at that time (interview April 19, 1997). At the second meeting the group started brainstorming about ideas for projects, varying from maintenance and extension of cycling-track to a symposium on the Shared Car, and a delivery service for shopping.

At the internal Work Conference in June 1996, the group presented the following projects:

- The Electrobus (presented by Van Rossum, Royal Dutch Transport, KNV);
- Transport-depot, demonstration bike (system of public bikes, presented by Luud Schimmelpennick);
- Car-sharing (Greenwheels; alderman Van der Putten will present a users pass to the first user in The Hague. A Greenwheels Car will be showed in the Town Hall);
- 'Fietstrommel' (Bike shelter; Company Wagemakers will show their Bike shelter in the Town Hall, see picture of a bike shelter below.)

To add a concrete contribution to the idea of shared car use, two members of T&T sought contact with an existing, small car share company called Greenwheels (Rotterdam). A representative of Greenwheels gave an introduction on the Conference, and the company also had an information stand in the Museon in The Hague, where the conference was held. Currently, three Greenwheels are in use, parked in a public garage. Working Group T&T further tries to extend the project of shared car use to a more complete "transport package," embracing a combination of a pass for a shared car, discount on Biesieklette and on public transport (HTM). The project with the shared car was successful, according to Creemers, because connections were made to existing initiatives, and because there appeared to be real possibilities in The Hague in terms of the demand side.

Whereas most working groups decided to stop the meetings after they finished (realizing) their *poekels*, T&T continued, but in a different form. Already a few times in the group's meetings it was mentioned that the group should be more broadly based, i.e. participants from the whole T&T field should be involved in the discussions on more sustainable transport in The Hague. Two active members of the group, Creemers and Apeldoorn, decided to invite specific representatives to the group, to continue these discussions after the formal HLA 21 was delivered. They try to have the meetings chaired by a representative from the City Council. They met in a building in the center of The Hague, called Oracle, and decided Oracle was a suitable name for their group as well, aiming at continuously coming up with new impulses in the process towards sustainable transport.

The first result of the Oracle group was the presentation of a Charter: "Sustainable Mobility Together'", in which the group managed to reach consensus on points of departure and possible solutions. The Charter was sent to political parties, executive departments of the municipality and of the region Haaglanden. The Oracle group tries to connect with existing plans for Traffic and Transport. In the meantime a representative of the Municipal Department of Traffic and Transport attends the meetings, as well as one from a Taxi company, from Biesieklette.

The results of LA21 are difficult to summarize. The project leader, when asked for the lessons learned, stressed the changed relations between citizens, companies and public groups and the local authorities. While formerly these groups were often acting against each other, a mode of mutual cooperation "to get things done" came into existence. The municipal administration is no longer merely seen as a delaying or restricting institution, but also as a partner helping to realize sustainable improvements. However, this is especially true for projects explicitly carried out within the HLA21. The actual influence of HLA21 on policy making is more limited, however. In The Hague the lay public, as well as public groups and private organizations, were all invited to participate in the building of the LA21. The project leader tried to reach a wide variety of citizens. However, the ones that showed up turned out to be mostly members of environmental groups or interest organizations: the broad public and the private sector were hard to reach. For many, however, the distinction between lay and representative was blurred. Even when people are (active) members of an environmental group, they can (and did) feel as if they were participating as a citizen and not as a "representative".

In the working group Traffic and Transport two active members regarded themselves as members of the "general public" (Creemers and Van Apeldoorn). But both have since become professionally involved: one works at Den Haag EcoStad and the other worked out the Charter for Sustainable Transport via his own private business. In the follow-up of HLA21 activities we see a tendency towards inviting people who are professionally involved in a specific field. In the Oracle Group, the follow up of Working Group Traffic and Transport, people participate on a personal basis, but come because of their professional interest. This seems to be influenced by the pragmatic character of this follow up. When it comes to realizing ideas and to more structural involvement, it demands more time and expertise.

The reasons why The Hague started to invite citizens to participate in the building of the HLA21, are summarized by Alderman Van der Putten: the creation of more public support for implementing environmental policy and the creation of possibilities for people to influence their own lives. The need for public participation is related to the "crisis" in local democracy that was recognized in the early 1990s and resulted in the process of Administrative Renewal. An important result of the HLA21 seems to be that various groups, formerly not cooperating, have been brought together. Although HLA21 embraces to a large extent representatives of public groups that were already active, new constellations have been made. Some participants (Creemers, Kuijpers) were very enthusiastic about these new contacts and links, which made it possible to implement their projects. Creemers: "Even as an individual you can realize your ideas and plans, when you manage to establish links with people who can help you with specific problems." In addition, the HLA resulted in more visible results, like bike stalls, a place for storks to breed, the planting of "eco-flowers" and shared cars. However, it is more the process of translation of ideas of the working groups into practice that is the most important result of the Local Agenda 21 in The Hague.

6. Conclusions

In general, we can discern two strategies in relation to LA21 in the municipalities that we have studied. The first is to try to create a "grass-roots movement", a strategy corresponding to the attempt at decentralizing Agenda 21, as well as the aim of engaging the citizens. The main efforts are geared towards education and information, which are in turn supposed to produce environmentally adjusted, thus sustainable, behavior among the citizens (especially in terms of consumption). The second strategy is to "adjust" the municipalities' activities in order to lead the way for other areas of activities and agents in the society, especially companies, and at the same time to accumulate knowledge regarding the transition to a more sustainable development.

We also want to emphasize the importance of the informal networks which are being constructed as a result of Agenda 21. Within the process itself contacts develop, most noticeable between environmental organizations and politicians and implementing agents, such as municipalities and authorities. The contacts are often mediated by the "professionalized" environmentalists working for various organizations or authorities. These agents make use of contacts they have developed during their activities with the environmental organization or environmental movement. They often remain a part of their organizations while at the same time they work for instance as Agenda 21 coordinators within the municipalities. This can be seen as different types of representation: on a formal level with the organizations and authorities, and in an informal level with network of contacts which has been developed in the broader environmental movement. It should also be noted that these strategies, developed locally regarding environmental issues, are used to drive the political system in a more environmentally adjusted direction (as in, for instance Växjö in Sweden).

Further, in terms of the work in Agenda 21 implemented in the municipalities, which individual person is responsible becomes a key factor. The individual's own network of contacts has been of major importance for how activities have been carried out and whether endeavors have been successful. If the Agenda 21 coordinators are part of a well-developed contact network among the various environment organizations and experts, they utilize these partly to access information, partly to influence local politicians. If, on the other hand, the coordinator is a novice in this kind of situation, for instance appointed as part of some sort of unemployment measure, or coming from within the municipal organization with a pronounced bureaucratic background, he/she often feels isolated from other agents out in society (not least environmental ones) and faces more difficulties when trying to access necessary knowledge and contacts. The position within the municipal organization is also of importance here. If Agenda 21 activities take place within the pre-existing municipal organization dealing with environmental issues, some gains can be made in terms of efficiency. There is already a developed organization with a certain network to utilize, but the work is secondary in relation to the current structures. If, on the other hand, the work is set up as a new organization within the municipality, as in Lund in Sweden, there is more independence in relation to the current structures, but the decision-process within the administration becomes problematic. It thus becomes a question of seeking a position which can be both independent and flexible, as well as have access to the established structure of power within the municipal administration.

References

Harrison, C. M., Burgess J. and Filius, P. (1996) 'Rationalising Environmental Responsibilities: A Comparison of Lay Publics in the UK and the Netherlands,' *Global Environmental Change*, 6: 215-234.

LGMB (1997) *Local Agenda 21 in the UK: The First Five Years*. Local Government Management Board.

Löfstedt, R. (1995) *The use of biomass energy in a rehional context: the case of Växjö Energi AB*, VEAB.

Macnaghten, P., Grove-White, R., Jacobs, M. and Wynne, B. (1995) *Public Perceptions and Sustainability in Lancashire: Indicators, Institutions, Participation*. Centre for the Study of Environmental Change, Lancaster University.

Macnaghten, P. and M. Jacobs (1997) 'Public identification with sustainable development: investigating cultural barriers to participation,' *Global Environmental Change*, 7: 1-20.

Tuxworth, B. (1996) 'From Environment to Sustainability: Surveys and Analysis of Local Agenda 21 Process Developments in UK Local Authorities,' *Local Environment*, 1: 277-297.

Veldboer, L (1996) *De inspraak voorbij – ervaringen van burgers en lokale bestuurders met nieuwe vormen van overleg*. Instituut voor Publiek en Politiek, Amsterdam.

Worcester, R. (1997) 'Public Opinion and the Environment,' in M. Jacobs (ed) *Greening the Millennium: The New Politics of the Environment*. Blackwell.

Young, S. (1997) 'Local Agenda 21: The Renewal of Local Democracy?,' in Jacobs.

Chapter Five

Roads to Sustainable Transportation?
On Public Engagement in Infrastructure Projects in Britain, Norway and the Netherlands

Patrick van Zwanenberg, Robbin te Velde, and Per Østby

1. Introduction

Most European countries are facing the same dilemma. On the one hand there is a pressing need for more effective transportation systems to keep the wheels of industry, trade and tourism rolling. On the other hand, environmental problems caused by transport are visible and increasing. In the EU report, *The Future Development of the Common Transport Policy,* subtitled "a global approach to the construction of a Community framework for sustainable mobility," it is stated that major environmental problems are caused by transport. A report made by the Norwegian Institute of Transport Economics (TØI), contains much the same problem definition: serious pollution of air and water, overuse of energy and free space, growing divisions between different regions, encroachment of the rural and urban landscapes, and negative effects on the natural environment and outdoor life (Berge et al. 1995). Throughout Europe, a new phrase has begun to work its way into the policy discourse: sustainable transport. But what roads (if any) are to be constructed in order to achieve sustainable transport?

One of the distinctive features of the kind of infrastructure projects that dominate transporation policies is the scarcity of *formal* mechanisms through which organised civic interests can participate in *core* aspects of the policy-making process. More often than not, civic influence is achieved, if at all, through the wider political domain, via actors and institutions other than those of central government, or in policy implementation rather than of core policy formation. One consequence is that the civic domain finds itself having to engage with pre-existing policy agendas, agendas that have been defined and framed by prevailing understandings about both the role and nature of policy and about what is actually at stake in any particular policy.

These official framings typically reduce what are complex technical, political and human problems to narrowly defined and tractable scientific and administrative issues, issues that are seemingly uniquely suited to expert and managerial control. That reductionist process tends to exclude or suppress less powerful and less well articulated human concerns. Furthermore, because the scientistic pretensions of the process are inconsistent with wider democratic control and deliberation, closure is effectively enforced around official framings that can and do conflict with civic concerns and understandings. In effect, whilst there may be channels that allow civic participation in some, non-core, aspects of the policy process these effectively require participants to conform to a set of non-negotiable bureaucratic rationalities and values.

In recent years, processes of State disengagement from several dimensions of the policy-making process are becoming evident across Europe as a consequence of the rapid spread of market values throughout the public sector – the privatization and deregulatory initiatives of the last decade or so – and the growth in importance of transnational forms of governance. In most countries, these shifts have acted so as to further disenfranchise the civic domain, since many crucial policy decisions have become less amenable to parliamentary oversight. At the same time, however, these shifts have engendered new opportunities for civic engagement, since the private sector and institutions such as the European Union play a more influential role in policy formation and thus become potentially more significant sites at which to attempt to engage with policy.

In relation to transportation, these processes are visible in different ways in different countries. This chapter recounts the experiences of civic engagement in infrastructure projects in Britain, Norway and the Netherlands.

2. Transport Policies in Britain

Transport has been one of the defining issues for the UK environmental movement. Since the early 1970s, environmentalists have highlighted the impacts of evolving transport trajectories on, for example, wildlife habitats, energy requirements, the use of other non-renewable materials, and a range of social disparities, cultural patterns of behaviour and quality of life issues. (see, for example, Adams 1981) More recently, the consequences for human health, ecosystem damage and global climate have also helped to provide a focus for a wide-ranging critique of the UK's escalating dependence on the more socially divisive and environmentally unsustainable forms of motorised transport. (see, for example, Roberts *et al* 1992)

By comparison with this broad critique, official conceptions of transport as an environmental issue have been extremely narrowly framed; for many years in terms of aesthetic impact, for example, such that policy responses were confined to planting trees along the edges of new roads. In the 1980s, some adverse physical effects of escalating private transport were grudgingly accepted and fiscal changes in favour of unleaded petrol and mandatory catalytic convertors on new cars were introduced; measures which originated externally and were initially vigorously opposed by the UK government.

Recently, the political salience of transport as an environmental issue has risen considerably. An erosion of the logic of current transport trajectories has become manifest with severe road traffic congestion. A large number of actors realised that this could not be curtailed merely by building new road infrastructure, or at least not unless building were to occur on an unprecedented and unacceptable scale. Furthermore, international commitments to stabilising carbon dioxide emissions and acute public expenditure constraints as the UK economy went into recession in the early 1990s have raised the political costs of even attempting to build out of congestion problems. On the back of those problems direct action and protest by new grass-roots organisations such as Earth First! and locally organised anti-roads groups, together with active campaigning efforts from the established national environmental groups, have helped to ensure that transport as an environmental problem has been firmly established, for the time being at least, as a mainstream political issue.

These recent developments have helped to open some space within the policy domain for deliberating a range of possible environment-related technological and policy options. Indeed several recent policy initiatives have an explicit environmental origin. These include, for example, new local planning guidelines that attempt to integrate land use planning and transport provision at a local level, fiscal policies on fuel, and a research programme investigating urban traffic control technologies and information technologies for transport users. (HMSO 1996a) As yet, however, there is no coherent environmental transport policy, and there continue to be widely divergent framings of what is at stake environmentally in the transport arena. Most of the policy options and technological research strategies deliberated by government are concerned primarily with improving the efficiency of existing transport patterns or, at best, with slowing down an increasing rate of environmental damage. For many environmentalist critics, such measures will do little to seriously challenge what is seen ultimately, by some, as a problem of unsustainable levels of mobility. (Adams 1992)

In the following sections we explore the principal opportunities for, experiences of, and tensions associated with public participation within one aspect of UK transport policy; namely, that of inter-urban road infrastructure. Two primary reasons underlie that focus. Firstly, the Department of Transport has tended historically to frame the transport problematique as one concerned predominantly, though not entirely, with roads. Exacerbating that tendency, over the last twenty years, have been wider political objectives which have resulted in the privatisation and deregulation of most of the transport industries with the exception thus far of road infrastructure which, traditionally, has been free at the point of use in the UK. Throughout the 1980s, for example, there were no "transport" white papers although "roads" papers were produced in four separate years. Thus, 'transport policy' in the UK, in the sense of a deliberative executive activity, is in many ways synonymous with "roads policy". Secondly, roads policy has provided the main focus for various types of formal and informal civic engagement in policy-making, not only on roads policy *per se* but also as an indirect means for engaging with transport policy more generally. The approach taken here is historical. We begin by describing roads policy and public representation in the 1970s and then move on to the period from 1979 to the early 1990s.

2.1 Roads Policy & Public Representation: 1950 - 1979

UK transport policy is made largely within a producerist policy community centred around the Department of Transport (DTp) (known before 1970 as the Ministry of Transport). The DTp has been and is currently responsible (within the broad policy agenda of government) for developing policy for all types of transport and it has executive responsibilities for the construction and maintenance of the inter-urban road network.

Producer interests historically have had, and continue to have, very substantial influence within the DTp and indeed government more generally. Representatives of the motor industry, road construction industry, oil industry, haulage industry and the motoring organisations (whose membership consists of the car driving public) constitute a powerful pro-roads "lobby" which, especially in the post-war period, has had a hugely significant role in the formulation and execution of government policy. (Finer 1958; Plowden 1970; Painter 1980; Hamer 1987) By contrast, environmental groups, and most other civic organisations, historically have had little effective purchase within the DTp. Some groups, such as Transport 2000 and Friends of the Earth, gained consultative status in the late 1970s, (Dudley & Richardson 1996b) although those and other environmental groups conceded during the early 1980s that their consultative relationships with development-orientated departments such as transport were largely token gestures. (Lowe and Goyder 1983, p. 64)

In the mid 1950s, with the lifting of post-war expenditure constraints, a rapid rise in private car ownership, and the concerted lobbying efforts of producer interests, considerable political momentum had built up behind long-existing plans to construct a strategic inter-urban road network. (Kay & Evans 1992) The then transport Minister, John Boyd-Carpenter appreciated the political advantages both in assisting the motoring public and in the promotion of an image of modernity and prosperity. (Dudley & Richardson 1996b) Boyd-Carpenter raised expenditure from the Treasury for a national road network and fostered the creation of an epistemic community of highway engineers within the then Ministry of Transport. Road engineers were sent to be trained in North America and Continental Europe and, over the next few years, professional and administrative hierarchies within the Ministry were integrated, thus upgrading the status and influence of road engineers in policy-making. (Painter 1980, p. 172; Dudley & Richardson 1996b, p. 574) By the end of the 1960s, regional Road Construction Units were established, responsible for design work, taking road schemes through the relevant statutory procedures, and acquiring land. These were staffed predominantly by highway engineers and run by road-building enthusiasts. (Levin 1979, p. 22)

The Ministry's success in obtaining firm government commitments to long-term targets for the road network – the 1970 White Paper *Roads for the Future* envisaged the creation of 3,500 miles of motorway and other inter-urban roads over a twenty year period – helped to secure a steady stream of funds from the Treasury. Road construction activities took place, and continue to take place, largely outside of any local authority or planning controls and were and are divorced from an integrated transport strategy. Other transport industries that had been nationalised after the war, such as the railways and some bus companies, were set up as self-administering public bodies, so that the main tasks of the Ministry of Transport remained with the administration and construction of the road system. Even between 1970 and 1976, when the Ministry of Transport was merged with the new Department of the Environment, the Highways Directorate was able to keep its unit intact and was treated as a separate entity by Ministers. (Painter 1980, p. 176)

In the late 1950s, the Ministry of Transport were aware that it would be advantageous for their negotiations with the Treasury if the economic case for new roads could be quantified, despite, at that time, a very favourable political climate for road building. (Kay & Evans 1992, p. 21) The Ministry's Road Research Laboratory produced an economic analysis of the proposed London Birmingham motorway in the late 1950s which justified the capital expenditure required by savings estimated in monetary terms, over a fixed time period, on the basis of journey times (including both working and non-working time), vehicle operating costs and reduced accident rates discounted back to present prices. This analysis was adopted in 1963 as the basis of a standardised cost-benefit assessment of individual road schemes known, since 1973, as COBA. The Ministry also began collecting data on traffic growth in the 1950s in order to justify the need for particular road schemes, and a system of measuring congestion was also devised. The Traffic and Accident Loss (TAL) measure used the results of timed car rides along with accident rates to measure levels of "loss" which, together with forecasts of vehicle ownership and use were used to arrive at a so-called 'overall picture of need'. (Painter 1980)

The 'scientific' COBA and TAL assessments effectively naturalised policy commitments to expanded road provision in which, it appeared, capacity demands would inexorably rise, and the economic returns to road building were high. In road traffic forecasting methods, for example, there is no recognition that there might be physical constraints on traffic growth. The forecasting equations implicitly assume that roads will be built to carry all forecast traffic. (Adams 1981, p. 165) COBA also has a self justifying logic since, for example, the monetary value of travel time saved for a particular road (travel time savings represent up to 80% of total benefits) is applied regardless of whether that road has the physical capacity to cope with the forecasts. (Bray 1995, p. 8) Similarly, the benefits offered by faster journeys were based on the assumption that a new road only causes a redistribution of existing traffic (thus ensuring faster projected journey times) rather than inducing additional traffic. Numerous additional assumptions inherent in the appraisal methodologies were and are of course extremely contentious, many of which are rooted in the highly selective (and surrogate) representations of value involved in the various approaches. For example, within COBA methodology, a series of less readily quantifiable social and environmental "costs" are ignored (such as the value of undeveloped landscapes or the social impacts of road schemes on local communities). The cumulative effect of the DTp's approach was both to create the notion that road construction always assists economic growth, even though the evidence for such a relationship is extre-mely poor, (Whitelegg 1994) and to portray road building as a technical and economic imperative rather than as a political choice.

The framing of those policy commitments within scientistic appraisal methodologies effectively rendered Ministerial and Parliamentary control minimal. Barbara Castle, Transport Minister between 1965 and 1968, noted, for example, that COBA seemed to have turned the roads programme into something based on technical formulae which Ministers could only rubber stamp. (Kay & Evans 1992, p. 43) Similarly, members of Parliament have noted that they were unable to understand the mathematical formulae used in the forecasting models. (Adams 1981, p. 147) In any case, Parliamentary control of the roads programme in the 1960s and early 1970s was virtually non-existent; indeed even Ministers had little effective control of the programme. Barbara Castle said that strategy on roads policy was effectively in the hands of the Ministry's road engineers, (Dudley & Richardson 1996a, p. 233) whilst her successor, Richard Marsh – Transport Minister from 1968 to 1969 – complained that his senior civil servants not only determined policy but did so in the interests of the pro-roads lobby, noting, for example, that "[t]he Director-General, Highways, who was to all intents and purposes Chairman of British Roads Ltd., was a Deputy Secretary in the Ministry of Transport" (cited in Dudley & Richardson 1996a, p. 222). As for Parliament, between 1966 and 1976, there was only one occasion in which there was a debate and vote on roads policy. Other important policy decisions were simply not discussed or voted upon in Parliament despite entailing massive commitments in expenditure. (Tyme 1978, p. 140) Furthermore, active attempts by Parliament to oversee transport policy during this period were not possible, as the 1975 Select Committee on Expenditure had revealed, since the DTp had failed to set out explicitly its intentions for an overall transport policy or any means by which Parliament could oversee particular investment decisions. (Tyme 1978, p. 140) In effect, roads policy was a product of administrative policy-making, constructed within the bureaucracy (in collaboration with private industrial interests) and shielded from effective Parliamentary, or wider public, scrutiny and oversight.

Yet there was one arena in which there did exist formal channels for public engagement in roads policy. Implementation of the roads programme falls under the land-use planning system. As such, there are statutory provisions, under the Highways Act, requiring the Minister of Transport to publish draft plans for a new or altered road (termed a line order) and await objections. If objections are unresolved and they have been made by statutory objectors, such as local authorities, then a public inquiry must be held. Where objections are received only from non-statutory objectors, for example, members of the public and many environmental NGOs, then a public inquiry may be held at the discretion of the Minister. Subsequent inquiries may also be held if there are objections to compulsory purchase orders on land and property adjoining the line order or to the construction of side roads. Following public inquiry recommendations, Ministers then make a decision about a scheme and objectors can also file for judicial review of a Minister's decisions on the grounds, for example, that the statutory powers within the Highways Act have been exceeded or not complied with.

In the early 1970s, local objectors to particular roads, which had hitherto been concerned primarily with shifting the line of a proposed road into someone else's backyard, (Dudley & Richardson 1996c, p. 18) were joined by national environmental groups. Both local and national groups began to take more strategic and adversarial positions at road inquiries, contesting, for example, overall transport policy and the long-term implications of unfolding transport trajectories on the physical and social environment. Objectors began demanding the right to expose the assumptions and political commitments implicit in expert traffic and road appraisal methods; in effect, trying to recapture a domain of political debate that had been appropriated by political interests either through scientistic techniques of factual "discovery" or as implicit unarticulated commitments that had never been subject to democratic deliberation.

The public have tended to view inquiries as a forum in which they will enjoy natural justice under rules of due process and neutral arbitration (as in a court of law). Yet, public inquiries have relatively restricted powers. The inquiry process is essentially an advisory mechanism in which Ministers appoint an Inspector to provide recommendations on a specific development. They involve an underlying presumption in favour of development and the wider policy framework around a specific proposal is exempt from questioning by objectors. Indeed they are supposed to aid implementation of preexisting policy and are designed to consider objections from *local* interests (for example, property holders) who may be affected by a *specific* development. In formal terms, policy is the responsibility of Parliament alone and thus public inquiries are concerned only with the *local application* of that policy. Furthermore, there are no statutory procedural rules governing the conduct of inquiries, and (as with other developments sponsored by, or closely associated with, government), Ministers are responsible for both promoting a development and, since Ministers are free to interpret inspectors recommendations as they wish, deciding on the merits of objections to that proposed development.

This mismatch between public expectation and the reality of the constrained role that inquiries actually have as a mechanism for civic engagement in the policy-process introduces an obvious tension. Objectors' attempts to use the vehicle of the inquiry system as a way of challenging wider policy issues, such as the need for a road, were frequently thwarted by Inspectors. Moreover, government could not be, and in the case of roads policy was visibly not, a neutral arbitrator, and this caused obvious problems for objectors expecting principles of natural justice to hold. Public inquiries also posed considerable problems for objectors because of the substantial time and resources needed to participate effectively.

Several additional characteristics of the inquiry process also proved problematic. Firstly, the distinction between policy and its local application rests on the constitutional doctrine of Ministerial accountability to Parliament, such that policy is a matter for Parliament and not self-selected participants at an inquiry. Yet, where policy originates from within the bureaucracy, as was clearly the case with roads policy in the 1960s and early 1970s, Parliamentary control is merely a fiction, thus contradicting the rational on which public inquiries rest.

Secondly, the distinction between policy and its local application is itself blurred. Given the massive resources required to build roads and the fact that existing road infrastructure shapes mobility patterns, each road development itself begins to constitute or affect policy. That lack of clear distinction was, and is, in practice, rendered more problematic by the tendency for the DTp to split inquiries into a new road, such that each inquiry deals with only a small part of a major new road. Each completed road section thus increases the overall policy logic for building the entire stretch of road.

Finally, the DTp's predictions of both traffic growth and the economic benefits of particular road schemes were justified by DTp representatives as government policy and therefore protected from criticism or discussion at public inquiries. But, the products of traffic appraisal methods, were validated as rational scientific claims, which would seem, in the eyes of objectors, to render those claims as open to deliberation and challenge. Yet in practice those "facts" were being substituted for political authority and as such immune from criticism at inquiries. (Wynne 1982)

The failure on the part of Parliament to control or oversee policy, the failure in practice of government to actually articulate explicit transport policy prior to or at inquiries, the negotiable distinction between policy and its application, and the naturalisation of policy commitments into a language of unchallengeable technical facts incensed many objectors. John Tyme, who represented the Conservation Society and other objectors at a series of roads inquiries in the 1970s, was, for example, convinced that the trunk roads programme was a 'consummate evil', the Department of Transport a "corruption of government", and the inquiry process a "grotesque denial of justice". (Tyme 1978, pp. 1, 92 & 101)

Because, however, road inquiries lacked prescribed procedural rules, Inspectors had a fair degree of discretion in determining procedure and deciding, for example, on what constituted admissible evidence. Objectors attempted to exploit that discretion and use the arena of the inquiry process to flag up a series of wider issues about both the roads programme and transport policy more generally. From the perspective of the DTp, such objections were seen as conspiratorial and as being propounded by subversives exploiting the inquiry process beyond its legitimate remit. (Kay & Evans 1992, p. 56) In response to the growing efforts of objectors to broaden the scope of inquiries, notes for the guidance of Inspectors at road inquiries issued in 1975 stated that a wide range of "policy" issues, for which Ministers were directly answerable to Parliament, were consequently inappropriate for consideration at local inquiries. These included, for example, the general assumptions made by Government about the availability and price of fuel, and the effects that these would have on traffic growth, issues of need, the objectives of the road programme, and the allocation of resources between different transport modes. In addition, the guidance notes suggested that inspectors should protect the Departments' representatives from cross-examination on the merits of those policy issues. (Levin 1979, p. 26)

Objectors nevertheless continued to challenge both issues of policy and the fairness of the inquiry process itself and began disrupting inquiries. Indeed, inquiries rapidly became a focus not just for conflicts over transport policy, but also about the legitimacy of the inquiry process as a whole and the authority of the State. John Tyme, for example, skilfully deployed techniques of civil disobedience in front of media audiences in an attempt to halt inquiries entirely. (Levin 1979, p. 28; Tyme 1978, ch. 1) At the Aire Valley inquiry in Yorkshire in 1975, for instance, the Inspector had the police remove objectors in order to continue the inquiry in public. Nevertheless, the inquiry had to be abandoned after protesters continued to demonstrate outside the inquiry hall. After the Aire Valley inquiry it became standard practice for the DoE to contact the police before an inquiry and have them ready to eject objectors if they insisted on raising procedural matters. (Morris 1976) By 1976, some roads inquiries ended in total disruption as objectors, in effect, turned the inquiries into political demonstrations. Where inquiries did get under way, it was sometimes only possible by conceding significant concessions to objectors over the remit of the inquiries.

The disruptions (given extensive media coverage) proved embarrassing to the government. Mindful of the political costs involved in the implementation of the roads programme and in the context of public expenditure crises in the wake of the 1973 oil price hike, the government decided to initiate a comprehensive review of transport policy. The review was steered by Bill Rodgers, Labour transport Minister from 1976 to 1979, who favoured the abandonment of the strategic roads network in favour of a more selective approach with greater consideration given to environmental factors. (Dudley & Richardson 1996b, p. 577) The review, published in 1977, admitted that there was no coherent national policy and promised to produce annual White papers on transport policy. (Levin 1979, p. 32) The White Paper also announced the abandonment of the strategic plan for inter-urban roads (despite opposition from officials in the DTp) in favour of schemes designed to solve local traffic problems. Expenditure on road construction had fallen significantly since 1974 as a direct consequence of the fiscal crisis after 1973 and the White Paper simply kept this at that lower level as a matter of policy. (Kay & Evans 1992, p. 57)

In the wake of the criticisms environmental groups had advanced at road inquiries Bill Rodgers also initiated a review of road appraisal methods. The review committee's 1978 report sanctioned some of the criticisms made by environmental groups claiming, for example, that the Department's traffic forecasts had a tendency to over predict traffic and that the cost-benefit procedures did not allow comparisons between alternative modes of transport. Moreover, the committee noted serious cases of bias in the calculation of COBA assessments. (Levin 1979, p. 33) One important outcome of the Leitch Committee was its recommendation that a Standing Advisory Committee on Trunk Road Assessment (SACTRA) should be appointed. This committee was established in 1978 and, notably, its membership included the Director of the Civic Trust, an environmental group, in addition to actors more closely associated with the dominant transport policy-making community. Finally, the government initiated a review of inquiry procedure, but although there were some changes to procedure (for example, over the provision of information prior to inquiries, the appointment of inspectors and the establishment of pre-inquiry procedural meetings), the Government nevertheless sought to diminish the discretion increasingly being exercised by inspectors and reemphasise inquiries' local role. These developments did, however, reduce the importance of public inquiry system as a mechanism for expressing public opposition to roads, partly because the pre-inquiry reforms allowed less scope for disruption at the beginning of inquiries. (Dudley & Richardson 1996c, p. 26)

A final blow to roads objectors occurred at the end of the decade when a judicial review of the M40/42 decision finally found in favour of the government after having been through the High Court, the Court of Appeal and finally the House of Lords. The legal action, Bushell v Secretary of State for the Environment, had been initiated by objectors on the grounds that a 1976 decision in favour of the motorway (following a public inquiry in 1973) had been invalidated by, amongst other things, the refusal to admit questions of need; in particular, objectors had not been allowed to challenge traffic forecasts and in the three years that had elapsed since the inquiry, several factors had affected the need for the road but the inquiry was not reopened. The High Court, in recognition of the administrative and political function of the inquiry process, rejected the case on the grounds that strict rules of evidence, that is rules of natural justice, did not apply at public inquiries. The case went to the Court of Appeal in 1979 which found in favour of the objectors on the grounds of natural justice. The Court claimed, for example, that the traffic forecasts, as facts, should be subjected to full examination and that therefore the public inquiry should be reopened in the light of new factual information. The Government then appealed to the House of Lords which, in 1980, decided that the Court of Appeal had been wrong and found in favour of the government. (Wynne 1982, p. 62) Thus in the end the courts had not only sanctioned the government's insistence that matters of policy should not be negotiated at public inquiries, but had chosen to uphold the use of "facts " – in this case traffic forecasts – as a substitute for political authority. In some respects, this judgment was a final blow to objectors seeking to challenge the spurious technicisation of politics so clearly apparent in the DTp's traffic appraisal methodology.

To summarise, from the 1950s to 1979, the civic domain had little opportunity to engage directly with the core transport policy-making community around the DTp, either directly or through Parliamentary, and to some extent even Ministerial, representation. The public inquiry process did, however, provide an arena for the environmental movement to not only criticise individual road schemes, (although only very few road projects were actually overturned through the inquiry system) but also road policy and transport policy more generally, through objectors' success in creating new political opportunity structures by pushing the remit of inquiries "upstream". As a consequence, environmentalist objectors successfully damaged the political credibility of the road building programme, and thus road construction would no longer be the obvious and uncontentious solution to increased car ownership and use. Objectors' activities also created a space in which expertise began to be demonopolised as actors from or representing the civic domain developed scientific and economic arguments challenging the "logic" of the roads programme. In addition, objectors began challenging the hegemonic position of the pro-roads transport policy community. Although not fatally wounded, by any means, small inroads were made by the environmental movement, as reflected, for example, in the appointment of an environmental member to SACTRA. And finally, objection and protests at public inquiries also provided a highly visible means through which to reach a sympathetic public and thus begin defining the domestic environmental agenda. (*cf* Grove-White 1991)

2.2 Roads Policy & Public Participation: 1979 - 1997

With the election of the 1979 Conservative administration, transport policy became dominated by a series of wider neoliberal political objectives. The market was imposed as the primary mechanism for determining transport modes, public expenditure was cut, transport industries were privatised, and policies were pursued that reduced the power of Trade Unions. Thus, planning deregulation and the abolition of the Greater London Council (GLC) and the Metropolitan County Councils meant, for example, that decision-making powers were shifted away from local authorities and local democratic processes to private developers and agencies of central government. (Blowers 1987) Bus deregulation and decreased subsidies to, and eventual privatisation of, the railways favoured the development of private forms of motorised transport. In 1981/82 public transport accounted for 16% of total capital expenditure on transport, but by 1988/89 this figure had declined to 2%. (Whitelegg 1989)

Although the strategic road network was more or less complete by the end of the 1970s, in the late 1980s, economic growth began increasing problems of congestion. Between 1979 and 1989, for example, the number of cars on the road increased by one-third and vehicle kilometres grew by 62% (Smith 1992). The roads lobby managed to persuade Conservative Ministers that new roads were required to cope with traffic growth and assist economic development. (Dudley & Richardson 1996b) National Road Traffic Forecasts issues in 1989 predicted that traffic demand could increase by between 83% and 142% by the year 2025 and in the same year the White Paper *Roads to Prosperity* announced a £12 billion roads building programme. At the same time, a Green Paper *New Roads by New Means; Bringing in Private Finance* announced that the government intended to extend privatisation from contracting for the construction of new roads to their promotion, finance and operation.

It shortly became clear, however, that growing congestion could not be solved simply by building new roads even on the scale envisaged in the Roads to Prosperity programme. (RCEP 1994) In several respects the warnings of the environmental movement in the 1970s were finally coming home to roost. The new M25 motorway – a ring road around London – filled up almost as soon as it had been completed, and many actors began to realise that there was no physical possibility of increasing road supply to a level that could meet forecast increases in traffic. This "new realism" amongst the transport policy community began to herald a shift in attention to the possibilities of demand management instead of the old policy of "predict and provide". (Goodwin *et al* 1991) furthermore, financial constraints on the road building programme increased markedly in the early 1990s as the UK economy went into recession. The roads programme was potentially a relatively painless way of reducing expenditure (at least in the short term). Indeed in 1994 the government announced a review of the roads programme, abandoning or postponing about 15% of its schemes.

At the same time as the review was announced, however, the Transport Minister, Brian Mahwhinney called for a "national debate" on transport in an attempt to reconcile economic and environmental objectives. Several factors played a role in that announcement. In particular, increasing congestion was occurring mainly in the south east of the country, an area almost entirely dominated by Conservative electoral constituencies and was obviously unpopular with a public putting up with the frustration of commuting to work. But, on the back of growing congestion, the activities of various different sections of the environmental movement were clearly significant in helping to reframe the public, and to some extent the official, image of the roads programme to something that encouraged traffic, caused adverse human and environmental health effects, threatened the countryside and dislocated communities. As an example of these activities, let us examine a proposal to complete the last section of the M3 Motorway through Twyford Down near Winchester in the south of England. The Motorway extension proposal had been subject to a twenty year campaign involving three public inquiries, as well as legal challenges under European and UK law. The extension was finally given the go-ahead in 1990 and by 1992, when construction began, the saga became the first focus for what later became a sustained programme of nationwide direct action after a range of grass-roots organisations occupied the site.

The plan for an extension of the M3 near Winchester went through two public inquiries in the 1970s. In 1971, at the first inquiry, plans for a motorway route across water meadows on the edge of the city were discussed but there was little organised opposition. By the time the second public inquiry was held in 1976, however, (concerned with compulsory purchase orders) the

Inspector was unable to get the proceedings started after uproar in the inquiry hall. Objectors refused to let proceedings begin, demanding that the earlier inquiry be reopened on the grounds that, at that time, the government did not possess an explicit national transport policy and thus how could there have been an "urgent need", as the DTp in 1971 had claimed, for the extension. (Tyme 1978, ch. 3) After five days of disruptions and protests the Inspector conceded that there was a case for looking at the need for the motorway and he adjourned the proceedings. At the reopened inquiry, several months later, the inspector recommended a reconsideration of the proposed route after objectors sought to show that the DTp's own COBA assessment for the motorway extension had a rate of return far lower than was normally required and that improvement to an existing road would be preferable. (Tyme 1978, p. 42; Bryant 1996, p. 11)

In 1983, new proposals were published by the DTp, but these continued to recommend a motorway extension, although this time instead of through the water meadows on the edge of the city, the extension was proposed on a route a few miles to the east of the city that would require a cutting through Twyford Down – an area of chalk upland containing rare habitat (designated as two Sites of Special Scientific Interest) and two scheduled ancient monuments. A public inquiry was held in 1985, but most of the objectors from the 1970s (principally local landowners) supported the scheme. Other objectors, largely local residents, argued, not about the need for the road, but instead suggested that a tunnel should be built through Twyford Down. That option was rejected but the objectors managed to reopen the inquiry two years later, on the grounds that several statutory bodies, namely, the Countryside Commission and English Heritage, had not given evidence at the inquiry – a shortcoming that, as far as the DTp were concerned, may well not have survived judicial review. (Bryant 1996, p. 70)

By the time the inquiry reopened in 1987, the objectors had managed to attract the support of local and national environmental groups, archeological societies, and the city council. The tunnel was costed at between 85 and 95 million pounds and was estimated to take 4 to 5 years to construct. According to the DTp, the COBA figures for the tunnel showed a negative return at – £21.35 million with low traffic growth as compared to the DTp's cutting route which was marginally negative at – £0.51 million with low traffic growth and COBA positive at + £8.99 million on high traffic growth assumptions. Thus, as far as the DTp were concerned, despite the damage to Twyford Down, only the costs and benefits quantifiable by the COBA assessment were valid in making a decision on the proposed route – which of course only included purchase costs for the land, thus excluding, for example, any ecological, amenity or historical "value" of the Downs. Indeed, given the logic of the COBA assessment, the more pristine and undeveloped a piece of land is, the more attractive it is to build roads on.

The objectors did not challenge or dispute the principle of the absence of environmental considerations in the COBA assessment since such a discussion would have been ruled out of order by the Inspector. (Bryant 1996, p. 117) Instead objectors argued that many other DTp schemes had a negative COBA, that the already completed sections of the M3 had been built at comparatively low cost, and that it was unreasonable that the final three mile link past Winchester should be self-supporting in COBA terms. Instead, any COBA analysis should reflect the cost of the M3 as a whole. But on the basis of the DTp's COBA assessment, the Inspector eventually concluded that he was unable to recommend the tunnel route. (Bryant 1996, p. 114) In 1990 the Government announced that the Twyford Down cutting would go ahead, as recommended by the Inspector.

Objectors then undertook a judicial review of the governments decision, in part, on the basis that the Government had failed to implement the EU Directive on Environmental Impact Assessment (EIA) for major construction projects into UK law. When the review found in favour of the government, the objectors then proceeded to appeal to the EU which began initiating an enforcement procedure against the UK government. The whole Twyford Down issue became increasingly high profile, in part due to the well-organised efforts of the objectors, the threat of EU legal action, but also in the context of the high public and media interest in the environment in general and roads in particular following the Prime Minister's conversion to the environment in 1988 and the announcement of the massive 1989 road building programme. Friends of the Earth (FoE) and the World Wildlife Fund had been assisting the Twyford campaign financially and in 1992, after the initial contract for construction had been awarded, FoE and local residents began short-lived occupations on parts of the site. They were subsequently joined by more militant protesters from Earth First! FoE withdrew, however, later that year, after injunctions were served on the group, but the protests continued with hundreds of people occupying the site, repeatedly clashing with police and security guards before the protesters were eventually evicted.

The whole Twyford Down saga illustrates well some of the tensions associated with the public inquiry system discussed previously. Firstly, the financial and resource costs of actually participating were enormous. The objectors in both decades happened to be extremely well connected and organised, and included architects, lawyers and engineers who could and did provide professional services. Even so, the costs, for example, of the judicial review were upwards of £75,000, and the consultants fees for the 1976 challenge to the DTp's COBA assessment were £40,000. (Bryant 1996) Secondly, the case highlights the exclusion of environmental values and the naturalisation of political commitments to road expansion, both of which were incorporated into a policy framework immune from challenge by dissenting expertise. The effective disenfranchisement of civic objectors entailed that protest, in both the 1970s and 1990s, was one of the few means to actually articulate civic sensibilities within the political process.

The Twyford Down protests led to the emergence of large numbers of anti-roads groups, generally independent of the established national environmental organisations. Within a year of Twyford Down, over 200 local anti-roads groups had been established, and umbrella organisations and networks such as ALARM-UK sprung up providing information and support to local groups. The new grass-roots organisations have tended to reject the State as a legitimate representative channel and have viewed the public inquiry system and other officially sanctioned means of engagement as a sham. Occupations and disruptions of numerous roads schemes have followed, including the blocking of existing roads in cities. In response to the direct-action activities of the anti-roads movement, the government attempted to criminalise occupations and site demonstrations through the 1994 Criminal Justice and Public Order Act which created a new offence of "aggravated trespass". The Act has served mainly to create further conflict and confrontation and has added an additional political target for the anti-roads groups.

Although the protests have had only limited success in stopping new roads, the financial costs (largely through extensive policing) involved in constructing new roads have been significant, as have the political costs associated with the suppression of social and environmental values held more widely than just with the individuals engaged in direct action at the sites of new roads. Indeed, a wide section of the electorate has been somewhat sympathetic to the protesters and in 1994, twelve backbench Conservative Members of Parliament whose constituencies were affected by the new round of road building set up an anti-roads Parliamentary Group. (Dudley & Richardson 1996a, p. 79) After 1992, the Government was re-elected on a narrow majority and thus securing approval of the roads programme consequently became more difficult.

In the wake of the government's 1989 roads programme, several of the large national environmental groups such as Greenpeace, Friends of the Earth and the Council for the Protection of Rural England relaunched major roads campaigns. In addition, large national groups such as the UK branch of the World Wildlife Fund, the Royal Society for Protection of Birds, and the Royal Society for Nature Conservation, began to engage with transport issues for the first time. Intergroup co-ordination through Transport 2000's Transport Activists' Roundtable (comprising some 20 different groups, including academics, consultants, professional organisations and government agencies such as the Country side Commission, as well as environmental NGOs) enabled the development of a broad environmental critique of transport policy and joint campaigning activities. (Rawcliffe 1994) These have included intense use of the media and use of the EU as a means of lobbying the UK government. As noted in the case of Twyford Down, the British environmental movement successfully enlisted the support of Carlo Ripa di Meana in 1991 over the failure of the UK to conduct Environmental Impact Assessments, not only over the Twyford saga but also over two other roads schemes (the East London River Crossing at Oxleas Wood and the M11 link road in East London). Though dropped after Ripa di Meana's resignation from the Commission in 1992, those events were of important political value for UK environmental groups. (Dudley & Richardson 1996a)

In addition, the emergence of an alternative epistemic community to that operating around the DTp, has been encouraged in large part by the national environmental NGOs. A Transport and Health Study Group was established, linking the environmental NGOs campaigning on transport issues with the more powerful health and medical lobbies. Reports published in 1994 by the Royal Commission on Environmental Pollution and the Parliamentary Transport Select Committee have, for example, highlighted the adverse health effects caused by vehicle pollution. (RCEP 1994) Other developments following the Rio Earth Summit in 1992 have also helped raise the political salience of transport as an environmental issue. For example, conflicts between the Department of Transport and the Department of the Environment have occurred as the latter became concerned that the government's targets for carbon dioxide reductions would not be met with existing policies on transport. And after 1992, the Prime Minister set up a Panel on Sustainable Development, which provided important and high level support for deliberating environment-related transport policies.

As yet, however, there are few indications of official policies that recognise and engage seriously with the wider environmental and social consequences of current transport trajectories. Moreover, the scope for government to direct transport policy is probably lower than at any other time in the post-war period. With the railway, bus and coach industries in the private sector, the government only has indirect influence over their investment and operating strategies. In the case of roads infrastructure the initiatives towards encouraging the private sector to finance and operate new roads may well hamper a more strategic environmental policy. For example, privately financed roads will require a higher rate of return than publicly financed construction, and because income for the private builders would be in the form of shadow tolls, which depends on total traffic volume, there may well be strong pressures to promote development along the route of new roads and ensure that competition from other transport modes is not promoted.

To conclude, whilst actors and organisations in the civic domain have rarely had any direct substantive influence in core policy-making processes, their activities illustrate various strategies for working around barriers to effective engagement. Environmental organisations have been adept at developing counter-expertise, utilising public inquiries and those institutions in which they do have access (such as the media, the medical profession and the European Union), as well as using protest and other less formal mechanisms of influence. Many environmental criticisms and objections to transport policy established in the 1970s have, as it were, come home to roost, and as such the role of the civic domain has to some extent been prophetic.

3. Sustainable Transport in Norway

In Norway, the transport sector has traditionally been regarded as closed to outsiders, while at the same time subject to a high level of public contention. In addition, not only huge financial resources, but also important social and environmental interests are tied to this sector. Inside this sector or segment, the cleavage between environmental and highway interests is much stronger than in any other sector (Klausen and Rommetvedt 1996: 153).

In a white paper from 1988, the problems caused by transport are stated as follows: "Nearly 300 000 persons are living in areas where air pollution is well above internationally recommended standards. Cars are the heaviest contributor to this situation, especially in our cities." (Miljø og utvikling 1989, p 95).

Nearly ten years later, one could read the following in a Parliamentary bill from the Ministry of Environment (MD): "In 1994 nearly 660,000 persons were exposed to NO_2 exceeding recommended values and 700,000 the values for PM10. There have been no evident changes since 1990." (Miljøverndepartementet 1995-96, pp 72-73)

The description gives a similar picture of the situation, but this time in a more detailed form, with indicators specifying the various components of air pollution. The capability of defining problems more 'accurately' are at hand, while the same capacity of defining suitable policies and solving the problems does not seem at hand at all.

The transport sector is without doubt a major contributor to air pollution, which in addition to the global warming, are causing illnesses such as asthma and cancer in the immediate environment. In Norway road transport, the fishing fleet and movable oil rigs produce about 37% of all inland emissions of CO_2. Concerning regional pollution, 22% of the total SO_2 and 66% of NO_x comes from transport. Road traffic is also the main contributor when it comes to local pollution, particular regarding NO_x, CO_2 and particles. In spite of a relative decrease of leaded gasoline and NOx, the increased traffic frequency still makes this sector, and cars, the very core of the problems.

The absence of a native automobile industry has not constrained the development of a Norwegian society heavily dependent on cars. If we compare with our Scandinavian neighbours, the density of cars in Norway in 1989 was 2.6 persons per private car. The corresponding figures for Sweden was 2.4, Denmark 3.2 and Finland 2.6. Even if Norway had a lower density than either USA with 1.7 or Germany with 2.1, it is safely placed in the upper level of car-dependent societies. Transport with cars, both for private purposes and for freight, has shown a steady and unquestionable increase throughout the whole post-war period. More interesting though, are the figures for the period after WCR. While the personnel transport by car rose by 20% from 1986 to 1996, the use of train, tram and bus (taken together) had no increase in the same period. When it comes to freight, there has been a decease in transport by train by 10%, while lorry freight has risen by 9% in the same period (Hille 1997).

The figures uncover a development that is far from what could be called sustainable. This development has been prolonged in "the regime of sustainability". The entrenchment of cars into all levels and sectors of society has created a situation that appears difficult to alter. Different concepts have been used to describe status quo. One way is to look at automobility as a system which has reached a high grade of maturity. Such concepts illustrate the main problem, the rigidity and the momentum, but leave out human agency and the free choice of people. People were not seduced or forced into a population of car-addicts. The present car-based transport system was not thrown upon us by some others. It was shaped and reshaped by users, businesses, experts and decision-makers. The same actors are presently unable to rebuild or reopen the system, and to create more sustainable transport solutions.

Even if many environmental problems must be linked to the widespread and intensive use of cars, the car represents not only a problem. It is sometimes too easy to forget that the private car offers a highly flexible, useful and democratic means of transport for the general public, industry and trade. In addition, the car is a desirable object in symbolic terms and as a commodity. Norwegian authorities are in other words confronted by the same dilemmas, trying to give form to its policies, as e.g. Swedish and French authorities. The car, as the core of transport, is both a problem and a benefit.

The origin and development of mass-motorisation has been analysed and described in various ways. It has been described in local and international terms, according to causes and effects, as an industry, in relation to social and cultural changes, in terms of transport policies, highway construction, the professions, and the transformations of cities. The list could be extended indefinitely. The dangerous and harmful effects of car transport are documented through a series of studies, and the main bulk of these studies deal with economical or technical aspects of transport and car use. In addition to an emphasis on technical and economic factors, many transport studies have focused on the functional and economical sides of car use and ownership.

The use of cars as well as studies of consequences have been explained according to factors such as time, costs, speed and queues. This rather narrow way of understanding the past, present and future situation has marked white papers as well as scientific reports. Recent studies have to some degree tended to move away from techno-economical considerations and towards socio-technical processes, offering new possibilities to approach processes that produce traffic. In spite of the deficiencies pointed at above, the emphasis on techno-economical considerations and the functional sides of car-transport, the produced insights have also resulted in the development of transportation and environmental policies and instruments.

A major part of these policies have had as their aim to reduce the private use of cars. To reach these goals the most common measures have been taxes on ownership, on sales, on petrol and on usage of road tolls. In addition, policies are increasingly being developed to reduce air pollution by setting emission standards. Presently the authorities are forcing the same type of pollution monitoring and control that previously has been used for industry. General taxation and more targeted taxation (such as toll roads) have brought more money to the Treasury, with no reduction in car use. One good example is the use of toll roads to the major cities. This system was introduced as a means to reduce the number of people entering the cities by private cars. At present the arguments are turned away from environmental benefits towards the worries of this income source getting too small to finance more road building.

One obvious conclusion is that the car in Norway has to a great extent been an abundant resource for the Treasury to get money for other important political goals. Even if there is an ensemble of policy instruments designed for transport, most of these instruments have not primarily been developed for environmental aims. In spite of this fact, some of them could have such effects. E.g. The Road Act is rather ambiguous concerning its environmental objectives. On the one hand it is meant to impose effective and rational transport solutions. On the other it can be used to reduce noise and pollution from transport. The Road Traffic Act has a more narrow objective, setting the standards for the amount of noise and exhaust allowed (Jansen and Osland 1997).

Presently, policies enforcing technical improvements have been more successful, than economical means. This statement has several implications. While the set up and use of economical regulations can be done by Norwegian authorities, the development of the technical features of cars are done by the international car industry. With only a limited share of the international car market, the ability to influence the manufacturers to produce cleaner cars is limited. On the other hand, standards presented by the pollution control agencies, in a very concrete way, set limitations to the design of the car. One more interesting solution has been the way in which heavily populated areas are designed and planned. The extensive use of speed-bumps, city-courtyards and 'zig-zagging' lanes reduce transport intensity in these areas.

In spite of its limited success, taxation is a frequent and increasingly utilised policy instrument. The taxation of leaded petrol, the CO_2 charge, the duty on diesel in addition to a variety of other taxes have had a strong growth in the last years. The heavy taxation has caused a series of debates, not so much because of their hitherto seemingly uselessness, but more related to questions of equality. While income taxes are progressive, transport taxation is flat. The paradoxical situation is illustrated by the fact that the same persons criticise taxation, bad roads, as well as the environmental burdens of car transport.

In addition there have been conflicts related to who shall have the incomes: the counties or the Treasury. Presently there has been a strike carried out by lorry owners to reduce the taxation. The strike has made the Parliament change their mind concerning a further increase in taxation. All in all the taxation has made it economically feasible to build even more and better roads in Norway, thus promoting road transport!

In addition to taxes and standards, there have been serious attempts to improve the position and use of collective transport services. Between 1991 and 1993 The Ministry of Transport (SD) and MD spent nearly 500 mill.

NOK on projects to improve the public supply of collective transport services. The projects caused a increased use of these services when pedestrians and bicyclist started to go by bus or city trains instead of walking or using their bikes. On the other hand there was no decrease in the use of private cars that could be related to the improved support of collective transport services (Hille et al. 1994: 35) The use of taxation, emission standards and the attempt to seduce drivers to go by bus instead of cars depicts the problems and the complexity of creating more sustainable transport solutions.

In Norway the construction of all types of infrastructure is both difficult and costly because of the natural barriers such as deep fjords and steep mountains. The construction of railway lines and roads demands substantial economic costs. This problem is added to by the spatial distribution of the Norwegian population. There are some large cities, but quite a large portion of the 4 million people in our country are scattered in relatively small rural communities with long distances between them. The continued existence of these communities is secured through a long line of political consensus. In addition there is a similar consensus regarding an equal supply of services to both the rural and urban society.

Concerning the rural part of Norway, the car is essential to keeping the small communities going. Financial restrictions therefore have a much more dramatic and harmful effect in the districts than in the cities, where one type of transport can be exchanged with another type with lower costs. These factors make changes in transport systems in the rural part of Norway very difficult.

On the other hand, the regional and local environmental consequences of transport are less acutely felt in the countryside. If we leave the global problems caused by transport aside, transport as a environmental problem is a city problem. Norwegian cities were to a large extent reshaped in the booming post-war years which also was the golden era of mass motorisation. They were therefore planned for people travelling with cars.

In addition a Norwegian tendency is to have detached houses. The one car and one house ideal appears to be an important factor in maintaining a high level of car use. This ideal has also had a strange effect on urban sprawl, and thus on car use. Between 1960 and 1990 the populated areas of the cities grew by 170%, while the population grew by only 27%. Since the cities, by European standards, are small, widespread and with low population density in the suburbs, it is difficult to set up collective transport services and to make these services efficient.

What are the possibilities for change? The communication/transport sector has a tradition for being very closed off and withstanding to external forces.

The sector have been dominated by two sorts of world-views or discourses. These are a techno-economic discourse and a land and areas planning discourse. While the techno-economical discourse was the answer from expertise to the early post-war problem, to construct a high-way system fit for the motor-age, the land and area planning discourse was a reflection and a reaction to the problems created by the outcomes of the first discourse.

These two types of expertise and their institutions express the two dominant discourses of the transport field in the post-war era. TØI represents the techno-economical discourse with emphasis on cost-effectiveness and calculation. NIBR represents the physical planning discourse with a focus on spatial planning, using The Planning and Building Act. While the first discourse, definitions and policies considered, works according to use of taxes and subsidies, the second operate in with spatial disposition and land and areas laws, and technical standards as its main elements.

In relation to the first discourse a differentiated taxation of transport according to modes of transport and environmental labels has been the chosen solution. This strategy has variations such as differentiated taxation on traffic "barriers" and on parking. This type of view on shaping more sustainable transport solutions in the future, has had some leeway. The present discussion in media are dominated by a new discussion of introducing a more extensive road pricing system.

The second discourse has recently gained better tools and more influence, through the modernisation of PBL, first in 1985 and later through the revision in 1993. PBL is primarily a tool for the municipality. A new means is the set up of maximum rates for emissions. They will take effect from 2002. In the last revisions of PBL two new elements are integrated: First the role of voluntary organisation as central parties of hearings, secondly a mandate for the central authorities to intervene in local affairs. According to the new line, the central administration can if necessary steer the development of previous undeveloped spaces in connection with tasks of high priority. One ambition for this new policy of intervention is to increase the population density in cities.

Both of these discourses may be termed technocratic for their heavy reliance on professional expertise, and their rather limited interaction with outsiders. Another important aspect is a kind of moving together, or more correctly, forcing together of the two discourses by the initiatives set up by central and local authorities. Even if this is a description of a development close in time, and at the moment is more theoretical than empirically supported, there seem to be a new way of defining problems and policies.

3.1 Two examples – TP10 and Green Cities

If we move from the more general viewpoints on transport policy, to present attempts to create more environmentally friendly transport policies, a gradual development from isolated transport solutions, to "whole city solutions" is taking place. Some actors, such as MD want to enforce on the sectored planning system more cross-sectorial and multidisciplinary planning attitudes. To illustrate this development we will look at two different projects: Environmentally Friendly Transport Plans for Larger Cities (TP10) and the Green City Project (MBP).

In 1989 the project Environmentally Friendly Transport Plans for Larger Cities (TP10) was initiated by MD and The Ministry of Communication (SD). Ten cities received money to create new and more co-ordinated plans for their local transport systems. The main argument for the initiative was increasing cost- and capacity problems in the largest cities. The terminology and the definitions depict the central role of technological and economical expertise defining problems. A secondary motivation, stated in the mandate, was the international focus on environmental problems, which illustrates MD's use of international signals to influence other sectors.

The mandate specified quite explicitly that TP10 should be carried out differently from earlier projects. Hitherto sectored task and responsibilities, such as the co-ordination of investments and the distribution and use of areas was to be co-ordinated. By bringing together different sorts of professions and including interest organisations and environmental organisations, all types of aspects related to the planning of roads should be considered. A final key difference between TP10 and earlier projects was that environmental considerations were to be treated as a premise, and not as a consequence. The participators were in addition to municipal politicians, the county officials (planning and public transport departments, the County Roads Offices and the County Environmental Departments.

The mandate of TP10 had elements of both the old and the new definition of problems. While the main argument for starting the projects a narrow definition of technical and economical bottlenecks in cities, the methods proposed were distinguished by a conception of complexity, cross-sectoriality and cross-disiplinarity. The forging of levels, positions and professions should in the view of the initiators create a more open planning process. In many ways this was an attempt to break open the transport sector. It was also an attempt to create a new constellation between previously remote actors, thus making weak links, strong ones. It was also, putting it a bit more bluntly, an effort to create new sorts of networks, and possibly weaken the existing ones.

TP10 was in addition to an attempt to formulate new policies, and an attempt to force new planning methods into local and regional administration.

The plan was evaluated in 1993. According to the evaluation and as a major conclusion, "the work seemed promising". Beneath the rhetoric the main conclusion was that the TP10 projects had not treated environmental consideration as a premise, but as a consequence. In other words, the "promising project" had failed when it came to one of its main aims. According to the evaluation, there were several reasons for this negative result: One was the link between TP10 and central and local highway programs. The projects were expected, correctly or not, to give input to these plans. This meant that projects that were to forge more sustainable transport solutions, were already tied up in the old system. A second problem was the old view that the improvement of roads would benefit the environment. In this view, more environmental friendly cities meant more environmentally friendly roads, which meant more roads outside the cities. More roads outside the cities meant less pollution in the inner cities, but more regional pollution. In addition better roads, meant better access by car, and in the long term, more transport and pollution. A third factor was the exclusion of collective transport services from the planning processes. To develop a system that did not relate car transport to collective transport was in all respect no step forward.

The evaluation also considered the type of expertise involved in the projects. In other words, had the attempt to create new networks succeeded? According to the evaluation, the main expertise had been engineers. Local politicians and interest organisations had participated only in a very limited way. In addition to their timid participation, they had neither had the power, nor the resources to affect the projects in a substantial way. To some extent, they also held views that did not support environmentalism. A report from TP10 in Drammen exposed serious problems and differences between the various types of expertise and participants. Especially stated were difficulties in getting the 'road-people' to accept that also environmental terms should be central in the planning. A last factor, especially underlined in the evaluation, was the connection between the investment in new roads and the financial situation of the municipalities. The primary aim of the local administration was to collect financial resources for to their regions. Supporting more sustainable transport was a question that was treated in separate terms.

In 1992 MD invited the 20 largest cities to apply for a project to develop sustainable municipalities and cities. The Green Cities Project (MBP) started in 1993, with Bergen, Tromsø, Fredrikstad, Kristiansand and one part of Oslo as participators in the project. MBP had, in the same way as TP10, as its main objective to integrate national environmental policies. In addition the mandate stated that a vital goal also was to develop new local policies. MBPs should reduce the use of undeveloped space, energy and resources. Furthermore MBPs should protect natural areas and green lungs, create "living" city centres and safer local communities. A central aim was to create environmentally friendly transport solutions.

While TP10 was a project for transport that was meant to draw on other sectors, MBPs were meant to deal with many sectors, and environmental tensions in cities in a broad manner. In this way it should also include transport as one of several elements influencing city development. To reach these goals each participating city was to develop local priorities for sustainable development. They were also to create local plans and concrete projects that could "visualise" long term goals. Again, the need for more cross-disciplinary initiatives was particulary stressed. It was also seen as important that the civil society, represented by voluntary organisations should be integrated in the work. According to the guidelines, new and alternative solutions should give inhabitants "a real possibility" to be environmentally friendly. Since the development of a "better" transport plans were to be central parts of the projects, transport planning was to be co-ordinated with the undeveloped space policy, which (as described above) had gone through substantial changes. This underlines the attempt to join the techno-economical discourse and the land and area planning discourse. A pamphlet from MD, the Ministry of Transport (SD) and SFT, illustrates both this attempt and the built-in tensions in the scenarios for greener cities. Some quotes underline these tensions: "The design of roads makes you a better motorist. Less car traffic means more effective commercial transport. The incomes increase. Soft road-users do have their own roads."

A brief look at one of these experiments might be useful. Old Oslo has traditionally been one of the worst city areas concerning pollution from car traffic. On this background it was picked to participate in the MBP. The inhabitants were invited to meetings, to carry out volunteer work and to be partners in hearings. While representatives from the general public joined in the start, their interest cooled, and the stage was left to experts and civil servants. The problem integrating the public illustrates our main theme. The concrete outcomes of the planning underlines another. One of the projects in Old Oslo was the construction of a tunnel for cars. When finalised, the tunnel contributed to substantially improving the situation for the local inhabitants. However, the overall picture tells of a substantial increase in traffic frequency because of easier access to the city. In other words, more cars, more traffic and more pollution, but not in Old Oslo. This problem, both connected to the question of more sustainable transport solutions and to the consequences of public participation, was addressed by professor Arvid Strand at the summarising conference of the MBP in 1995.

Strand stated the fact that there never had been a more intensive construction of new highways in Norway than today. Is there, he asked, a solution to problems, to move all transport under mountains? Strand ended his attack by stating that sustainability now had gained the same meaning as "love, babies and Christmas".

4. The Landscape of Dutch Infrastructure Projects

The Dutch have a strong tradition of large infrastructure works. The stories of the battle against the ever threatening sea are legend. This heroic image is still vivid today and is nurtured by a powerful and closely knit community of civil engineers. The professionalization of Dutch engineers dates back from the 17th century. About 1850 both the Delft University of Technology and the Royal Dutch Society of Engineers were founded. These institutions were from the start heavily populated by civil engineers. Through today, 'civil engineering' is synonymous for 'Delft'. The heroic image of the engineer at war against the dark forces of nature got a strong upsurge during the first huge infrastructure project: the land reclamation projects of the Zuyder Zee, just before and after World War II. In 1953 large parts of the southwestern parts of Holland were flooded. This set the agenda for the second megaproject: the flood barriers of the Delta works. The project established the central role of Public Works RWS), a semi-autonomous sub-department of the Ministry of Transport and Communication (V&W). Many young engineers who were involved in the Delta project hold today high positions within industry, universities and government. There are several big engineering and construction firms in The Netherlands who are major players in the international market. It is said that the spin-off of the Delta works (expertise building, public relations) has been very important for these firms. Since then, numerous infrastructure projects have been completed in foreign countries.

The career of Hans Smits is exemplary. He started his career as an engineer with the Delta project. Afterwards he became one of the top advisors of the Minister of Transport and Communication. His present function is president of Schiphol airport. Furthermore, he has been actively involved in the decision-making process of the Betuwelijn (the eastwards freight line which connects Rotterdam harbour with the industrial hinterland in Germany) and was one of the architects of the HSL (the high speed which connects Amsterdam/Schiphol and Rotterdam to the French TGV system).

In May 1997, the last flood barrier (in the New Waterway at Rotterdam) was put into use. Officially, the Delta works are now completed. To safeguard a third wave of large infrastructure projects, the network of the Delft engineers has been institutionalized into a formal organization, 'Nederland Distributieland' (NDL) which has been very influential. Four large infra-structure projects are now set on the political agenda or are already in progress: the Betuwelijn, the HSL, the expansion of Schiphol airport and the 'Tweede Maasvlakte,' a land reclamation project which will extend the industrial area of Rotterdam harbour further westwards into the North Sea.

The Delta network is still a stronghold within the Dutch policy on large infrastructure projects. However, the social and political circumstances have radically changed since the early 1970s. The report of the Club of Rome started a lively discussion on the quality of life. Nowhere was the "limits to growth" thesis so widely discussed as in The Netherlands. Environmental groups sprung up which were critical to technology in general. At the same time, a separate Ministry for Environmental Affairs (and Public Health) was founded. Public awareness on environmental issues grew and the centralized top-down structure of the decision-making process on infrastructure projects was criticized. One result was that the design of the Delta works was partly changed. The permanent flood barrier in the Oosterscheldt was made semi-permanent to retain the salt-water biotope. The success of the environmental groups (who tried to minimize the impact of the flood barrier on the existing nature) could be partly explained by the technical challenge it gave to the civil engineers (a semi-permanent barrier was a novelty).

At the end of the 1970s, RWS and its constituency and the environmental movement literally clashed in the forest of Amelisweerd. Here, the public image of the "heroic engineer" was severely damaged and RWS realized that institutionalizing public participation could diffuse or even forestall violent resistance. For the environmental groups, which had in general started as protest movements, Amelisweerd turned out to be a watershed. Many hard-liners were disappointed and left the environmental scene. The less radical activists adopted a strategy of dialogue. As a result of both the strategic changes of the civil engineers and the environmentalists, the 'inspraakpro-cedure' (or public dialogue procedure) became an important policy instrument.

At about the same time, nuclear energy was an important and controversial topic. In response to the political turmoil about the nuclear energy programme, the Dutch government decided to establish a nationwide "Societal Discussion on Energy Policy" (MDE), a kind of extended public referendum on technology policy options with regard to the official energy policy. Some critics consider the MDE to be a failure because of its limited impact on the policy making process. However, the MDE did influence the relations between the environmental groups, the Dutch government, and the energy-sector. It resulted in a more co-operative attitude on the part of the environmental movement towards the established political system.

During the 1980s, the dialogue between the (national) environmental groups and the engineers became further institutionalized, and several discussion platforms were created. Even normally action-oriented organizations such as Greenpeace became more and more involved in the formal decision-making processes. Most of the national environmental groups adopted a pragmatic strategy. Actions were limited to the local level and mainly aimed at increasing public awareness, and mobilizing public support ('achterban') for or against specific projects.

At the end of the 1980s, several large infrastructure projects appeared on the political agenda. During the decision-making processes on these projects, the efficiency problem of the "inspraakprocedure" came to the fore. The engineers pointed to the time-consuming character of the participation process. On the other hand, while there are indeed many ways for individuals or pressure groups to stall or delay the decision-making processes, this is often the only way influence can be exerted. The most important decisions are still made outside the public realm. Crucial design parameters were already fixed during the planning phase, which was not open to the public. Many environmental groups have come to believe that the 'inspraakprocedure' is often only window-dressing. In 1994, an important white paper was published by the Scientific Council for Public Policy (WRR). In the report, another decision-making procedure was suggested that deals with both the problem of speed and participation. A core concept in the new procedure is the *nut/ nood-zaak-discussie*: a public consultation at a very early stage about the fundamental question of the desirability of the project (i e, whether the project should be started at all). During the later stages of the decision-making process the elaboration of details should be left to experts (and not to the Parliament).

RWS was also looking for ways to deal with the problems of speed and participation. In 1994 it established Infralab, a special division. Infralab had to work on solutions for both the lack of support for infrastructural projects and the long procedures for decision-making on large projects. Its mission is to overcome the gap between the authorities, experts, and society. The working method of Infralab is to find concrete ways of organizing dialogue between RWS and other groups in society, thus bringing more creativity into the planning phase and shortening the process of decision-making. The working method starts – and this is a new component – with defining the problem together with users. In order to "take care of an open dialogue", Infralab treats participants as persons, instead of representatives of organizations. Traditionally decisions are prepared within RWS, a few alternatives are worked out, and thereafter the head of department signs for alternative *a*, *b* or *c*. The second and third step of the working method of Infralab is also different from the traditional procedures. From the problem definition of users, experts are invited to come up with solutions and in the third step possible actions are defined. The maximum time scale of the whole process is one year. Very important in the long run in the work of Infralab is the transition within RWS: from building projects towards organizing the process, bringing experts and users on to speaking terms again. Though this process is only starting, according to the head of Infralab, it has the support of the top of the department.

During the last decade, five large infrastructure projects have been started. The decision-making process on the expansion of Rotterdam harbour and the Second National Airport are still in their initial stages and have not been studied in detail yet. The other three decision-making processes will be reported on in the following paragraphs. To a certain extend these processes ran parallel and the outcomes have mutually influenced each other. Still, a certain chronological order can be distinguished (see table 1)

I. **Betuwelijn**

The construction of a freight line which connects the Rotterdam harbour to the German hinterland. The railway follows the Betuwe river basin.

II. **Schiphol**

The expansion of Schiphol airport. The contruction of an extra (fifth) runway.

III. **Hoge snelheidslijn (HSL)**

The construction of a passenger bullet train which connects Amsterdam/Schiphol and Rotterdam to the French high speed train railway network.

IV. **Tweede Maasvlakte**

A land-reclamation project to extend the industrial area of Rotterdam harbour westward into the North Sea.

V. **Tweede Nationale Luchthaven (TNLI)**

A national debate on the future Dutch aviation infrastructure (TNLI) structured along the Infralab-method.

Table 1. Recent Large Infrastructure Projects in The Netherlands

The environmental groups have characterized the decision-making process on the Betuwelijn as a deal between the Dutch Railways (NS) and the national government. Because both the NS cargo division and the ECT (a conglomerate of container handlers from Rotterdam harbour) needed the Betuwelijn badly, there has been a strong lobby from the NDL organization to push through the project. No *nut/noodzaak-discussie* has been implemented in the decision-making process. The NS, which dominated the project team in the early stages, presented the Betuwelijn as a *fait accompli* to the local governments. Irritation ran high. In some cities, local governments and local pressure groups took joint action. Meanwhile, RWS slowly took over the mediation process from NS because it realized that the project team and the public (including the local officials) had to be brought on speaking terms again. The change of government after the elections of 1994 gave the local and regional lobby coalitions a chance to influence the decision-making process on the project. A special committee was appointed to deal with the demands of the local government and environmental groups. Nevertheless, the protest resulted only in minor changes in the route. Tunneling, for instance, was banned by the engineers because it was considered to be too costly and technically too complicated. As for the environmental dimension of the project, the committee proposed some radical complementary policy measures to improve the position of rail transport vis-à-vis road transport but these have not been taken over by the Minister of V&W. There was however another important result of the Betuwe project: from now on, the *nut/noodzaak-discussie* had to be part of the decision-making process on large infrastructure projects.

There is a strong overlap between the decision-making processes of the Schiphol and HSL projects. The advocates of the expansion of Schiphol, for instance, have embedded the construction of the HSL at an early stage in their "mainport" project. On the other hand, in the decision-making process on the HSL project, an alternative route which did not include a stop over in Schiphol was never considered seriously. A *nut/noodzaak-discussie* has been implemented during the decision-making process on the expansion of Schiphol airport. However, as in the case of the HSL project, closure on important design parameters had already occurred before the 'nut/noodzaak-discussie' was started. The core assumption that Schiphol airport had to become a 'mainport' could not be questioned. Before any public participation had occurred, the mainport scenario had already been been worked out in detail by the national government (The Ministries of V&W, VROM, Economic Affairs), the regional and local government, the Schiphol Corporation, Royal Dutch KLM, and Dutch Railways. The environmental movement organizationss then tried to use the formal stage of public particicipation, in which the exact location of the additional fifth runway was decided upon, to block further growth (a sixth runway) of Schiphol airport. However, the option of a parallel runway, which was favored by the advocates of the non-zero growth scenario for Schiphol, eventually was chosen. The mainport concept stands for the core assumption of the present government policy, that growth in (air) traffic (or transportation) means economic progress and thus leads to a raise of income and the creation of new jobs.

At about the same time, the HSL project was started. Based on the substitution argument (high speed trains could substitute for a significant part of intra continental flights) the desirability of the HSL was never seriously questioned during the *nut/noodzaak-discussie*. The environmental movement, including the Green party in the Parliament, strongly supported the substitution effect. In the case of the Betuwe project, there had been resistance from the regional and local government. Instead, the HSL project had been put on the political agenda by both the regional governments of Noordholland and Zuidholland and the NS. Once again, closure on important design parameters had already occurred before the *nut/noodzaak-discussie* was started. Minimizing of the traveling time required a maximum speed of 300 km/h and the shortest possible connection between Amsterdam/Schiphol and Rotterdam. Eventually, this excluded all but the one route which had been originally proposed by the project team (the A1-option). Contrary to the decision-making process on the Betuwelijn, different alternatives for the route have been suggested. The social construction of the 'Groene Hart' by a wide range of professional organizations, pressure groups and the environ-mental movement literally blocked the A1-option. The Minister of VROM adopted the *Groene Hart* and so did a majority of the Parliament. The clash between the Minister of V&W (who still stuck to the A1-option through the 'Groene Hart') and the Minister of VROM resulted in a compromise which had not been possible during the decision-making process of the Betuwelijn: the tunneling of the *Groene Hart*. Still, a majority of the Parliament, backed up by the powerful public coalition, favored an option which circumvented the *Groene Hart*. The cabinet however was able to push through the A1-option by making use of the dissension in the Parliament. Both the public and the Parliament were very disappointed and felt betrayed. As a consequence, the cabinet realized that the *nut/noodzaak-discussie* had to be structured as a really open dialogue. In a next project (the expansion of Rotterdam harbour), it could neither bypass the Parliament nor neglect the public opinion. Furthermore, it had to take up the substitution argument seriously, which had played such a crucial role during the *nut/noodzaak-discussie* of the HSL project. The government could not, at least officially, promote both the railway transport and the air transport at the same time. This had a significant impact on the decision-making process on the future Dutch aviation infrastrastructure, which is in progress right now. The basic question which underlays that debate is whether to expand Schiphol even further or to build a new airport on another location (the most favourable option is the construction of an artificial island in the North Sea).

It seems that several trends are cumulating in the Tweede Maasvlakte project, the expansion of Rotterdam harbour. First of all, the cognitive space within the policy-making arena on large infrastructure projects has changed. The *nut/noodzaak-discussie*, which might have been mere window-dressing, has over time gained more importance. As for the social networks, in every subsequent project, more people and groups have been included in the decision-making process who do not belong to the technocratic network. In the 'Tweede Maasvlakte' project, professional organizations, pressure groups, the environmental movement and individuals have formally been included in the planning process. Thus, for the first time, the fundamental question whether to build the project at all, is seriously being addressed.

5. Conclusions

In Britain, Norway, and the Netherlands – as, for that matter, in the other PESTO countries as well – transport policies remain one of the most difficult areas to affect and influence by broader public interests.

As we have seen this is partly because public concerns about the environmental impact of transport are so difficult to square with current transport trajectories. Many of the policies that are deployed to help to reduce or limit environmental damage from private motorised transport, for instance, either shift the problems elsewhere or are simply outstripped by increases in traffic. It seems that any serious attempt to move towards sustainable transport systems is going to involve challenges to core assumptions about the desirability and inevitability of increasing mobility, the value placed on environmental and aesthetic goods and so on.

Yet, in most of the PESTO countries, transport policy is a technocratic endeavour in which such assumptions are rarely open for negotiation.

That is why the Infralab innovation in the Netherlands is an interesting "light in the darkness". As one of the few recent policy mechanisms to develop out of the interaction between the public and the dominant policy making interests, it represents a significant accomplishment. And, as the experiences show, it also indicates that public participation, in certain contexts, can improve the planning and policy process.

Nevertheless, whether, in the long run, effective forms of public engagement are merely a democratic achievement, or whether they can also help to move transport systems in a more sustainable direction, is still an open question.

Note
[1] (NDL): "The Netherlands Distribution Country", to stress the importance of transportation for the Dutch economy. NDL has a yearly budget of Nfl. 5 million. The Ministry of Transport and Communication and the Ministry of Economic Affairs contribute Nfl. 2 million. The participating firms contribute the remaining Nfl. 3 million. The NDL network consists of more than 650 high officials, captains of industry and politicians.

References

Adams, J. (1981) *Transport Planning: Vision and Practice*, Routledge.

Adams, J. (1992) 'Towards a Sustainable Transport Policy', in J. Roberts et al (eds) *Travel Sickness: The Need for a Sustainable Transport Policy in Britain*, Lawrence & Wishart.

Berge, G. et al (1995) *Bærekraftig og miljøtilpasset transport – noen ulike definisjoner*, Arbeidsdokument TP/0925/95, 0-2135, Oslo: TØI.

Blowers (1987) 'Transition or Transformation? - Environmental Policy Under Thatcher', *Public Administration*, Vol. 65, pp. 277-294.

Bray, J. (1995) *Spend, Spend, Spend: How the Department of Transport Wastes Money and Mismanages the Roads Programme*, Transport 2000.

Bryant, B. (1996) *Twyford Down: Roads, Campaigning and Environmental Law*, E. Spon.

DTp (1987) *Transport and the Environment*, HMSO.

Dudley, G & Richardson, J. (1996a) 'Why Does Policy Change Over Time? Adversarial Policy Communities, Alternative Policy Arenas, and British Trunk Roads Policy 1945-95', in *Journal of European Public Policy*, Vol. 3, No. 1, pp. 63-83.

Dudley, G & Richardson, J. (1996b) 'Promiscuous and Celibate Ministerial Styles: Policy Change, Policy Networks and British Roads Policy', in *Parliamentary Affairs*, Vol. 49, No. 4, pp. 566-583.

Dudley, G & Richardson, J. (1996c) *Arenas as Agents of Policy Change: New Institutionalism and the Public Inquiry Process in UK Trunk Roads Policy*, Essex Papers in Politics and Government, No. 104, University of Essex.

ESRC (1996) *Directory of Clean Technology Research*, HMSO.
Finer, S. E. (1958) 'Transport Interests and the Roads Lobby', in *Political Quarterly*, Vol. 29, pp. 47-58.
FoE (1994) *Roads to Ruin: Friends of the Earth's Response to the Roads Review in England*, Friends of the Earth.
Goodwin, P. et al (1991) *Transport: the New Realism*, Transport Studies Unit.
Grove-White, R. (1991) *The UK's Environmental Movement and UK Political Culture*, Report to EURES, November 1991, Centre for the Study of Environmental Change, Lancaster University.
Hamer, M. (1987) *Wheels Within Wheels: A Study of the Road Lobby*, Routledge & Keegan Paul.
Hille, J. et al (1994) *Redusert forbruk – kommunal handling: En idekatalog med eksempler fra Norden, Nederland og Tyskland*, Prosjekt Alternativ Framtid.
Hille, J. (1997) *Den alternative Nasjonalrapporten om Norges oppfølging av Brundtlandkommisjonen og Agenda 21*, FIVH.
HMSO (1996a) *Transport: The Way Forward; The Government's Response to the Transport Debate*, Cm 3234, HMSO.
Jansen, A. og O. Osland (1996) 'Norway' in P. Christiansen (ed),: *Governing the Environment: Politics, Policy and Organisation in the Nordic Countries*, Nordic Council of Ministers.
Kay & Evans (1992) *Where Motor-Car is Master: How the Department of Transport Became Bewitched by Roads*, Council for the Protection of Rural England.
Klausen, J. and H. Rommetvedt (eds) *Miljøpolitikk, organisjonene, Stortinget og forvaltningen*, Tane Aschehoug.
Laffin, M. (1986) *Professionalism and Policy: The Role of the Professions in the Central-Local Government Relationship*, Gower, Aldershot.
Levin, P. H. (1979) 'Highway Inquiries: A Study in Governmental Responsiveness', in *Public Administration*, Vol. 57, pp. 21-50.
Lowe, P & Goyer, J. (1983) *Environmental Groups in Politics*, Allen & Unwin.
Miljø og utvikling 1989 *Norges oppfølging av verdenskommisjonens rapport*, 1989, page 95.
Miljøverndepartementet (1995-96) *Parliamentary proposition*. No. 1.
Morris, T. (1976) The Highway Men, *New Society*, 9th September 1976, pp. 549-550
Painter, M. (1980) 'Whitehall and Roads: A Case Study of Sectoral Politics', *Policy and Politics*, Vol. 8, No. 2, pp. 163-186.

Plowden, W. (1970) 'MPs and the Roads Lobby', in A. Barker & M. Rush (eds) *The Member of Parliament and His Information*, Allen & Unwin.

POST (1995) *Transport: Some Issues In Sustainability*, Parliamentary Office of Science and Technology.

Rappert, B. (1995) 'Shifting Notions of Accountability in Public and Private-Sector Research in the UK: Some Central Concerns, *Science and Public Policy*, Vol. 22, No. 6, pp. 383-390

Rawcliffe, P. (1994) *Swimming With the Tide: The Changing Nature of National Environmental Pressure Groups in the UK 1984-94*, Unpublished PhD Thesis, University of East Anglia.

RCEP (1994) *Transport and the Environment*, Royal Commission on Environmental Pollution, 18th Report, Oxford University Press.

Roberts, J. et al (eds)*Travel Sickness: The Need for a Sustainable Transport Policy in Britain*, Lawrence & Wishart.

Smith, J. (1992) 'Roads in the 1990s: Expansion or Restraint?', in *Geography*, Vol. 77, No. 334, pp. 73-76.

Tyme, J. (1978) *Motorways Versus Democracy*, Macmillan.

Walton, W. (1996) 'Policy Changes in the Government's Road Building Programme: A U-Turn or Just an Application of the Brakes?' in *Town Planning Review*, Vol. 67, No. 4, pp. 437-455.

Whitelegg, J. (1989) 'Transport Policy: Off the Rails?' in J. Mohan (ed) *The Political Geography of Contemporary Britain*, Macmillan.

Whitelegg, J. (1994) *Roads, Jobs and the Economy*, Greenpeace.

Wynne, B. (1982) *Rationality and Ritual: The Windscale Inquiry and Nuclear Decisions in Britain*, BSHS, Chalfont St Giles.

Chapter Six

Environmental S&T Policy in One Country: The Public-Policy Interface in Italy

by Marco Giuliani

1. Introduction

The "greening" of political and policy processes has become a widespread phenomenon. By now, the path towards what might be called "post-environmentalism" – the integration of environmental concern in both public and private decisions – has come to be pursued by almost all the industrialized countries.

Whereas this uniform drift has been promoted by a set of extra-national factors – similar environmental problems, globalization of markets, deliberate efforts from international organizations, diffusion of ideas inside the scientific community, explicit pressures and campaigns by green associations and citizens – still the process has been influenced by national legacies and related to internal, domestic needs and agendas. Hence, the persistence of significant heterogeneities between countries comes as no surprise.

Notably, national experiences in S & T policy can depart from the ideal process that has led from a segmented to an integrated approach to environmental problems, in quite a few aspects:

1 in the timing of the different phases, as well as in the pace of the whole process, as it could be easily followed through a comparative analysis of the public institution building in this particular issue arena (Janicke 1991).

2 in the prevailing policy style adopted (Richardson 1982), particularly regarding the confrontational or consensual attitude exhibited towards non-institutional actors (private interest groups, environmental associations, firms, citizens, etc.).

3   and, finally, in the actual outcome of that process, that is in the institutionalization of environmental concern across different policy sectors, and in the implementation of public and private decisions.

This paper deals with the particular way in which these centripetal and centrifugal forces have combined in the Italian environmental policy-making. Specifically, it addresses the question of how different actors – politicians, bureaucrats, entrepreneurs, experts, environmentalists, citizens – interact in the redefinition of acceptable answers to environmental problems. In so doing, we will explore how the four domains which have been considered to frame the environmental S & T arena – that is, the bureaucratic, the economic, the academic and the civic – have coped with these problems, and how they have faced the cultural tensions which may have emerged during this redifinition process due to their different languages, goals and ethos (for the discussion of policy domains, see Jamison and Østby 1997).

In the paper we will argue that the Italian case presents an interesting paradox. On the one hand, it exhibits an evident comparative delay in the overall framing of the policy discourse; the absence of specific institutions, a relative lack of resources, insufficient transparency of the processes, limited participation and conflictual relationships have been common features during the last decades. Albeit widely shared by many analysts (Berrini 1994, Liberatore and Lewanski 1990, Signorino 1996, Lewanski 1997), this severe judgment of backwardness, however, has to be fine-tuned by a closer look.

Leaving aside the macro level of analysis, which still portrays the most apparent and palpable features of the Italian policy-making in this area, and shifting to a micro and local level, it is possible to find some interesting "niches" of excellence, a few positive experiences in applied post-environmentalism: municipalities which have been praised for their Local Agenda 21; private firms which have anticipated eco-audit procedures in spite of the absence of the Italian legal framework incorporating the EU directive; companies which have decided to bet on environmental innovation as a way to compete in the market; entrepreneurial associations which, in strict co-operation with university institutes, have been chosen by the EU to conduct pilot programmes of the Eco Management and Audit Scheme through their associates; environmental groups which regularly make up for public institutions in providing up-to-date information on the state of the environment; regional governments which lead the way in introducing new environmental technologies, and in experimenting with ways of communicating their plans to the citizenship and in maintaining international links notwithstanding (or against) the central passivity.

These islands of excellence can be considered in two different ways. They can either be thought of as the classic exceptions to the general rule – spotty outcomes of unpredictable contingencies – or regarded as interesting paradoxes worth explaining. In a normative perspective, it could even be argued that these experiences, showing positive outcomes in an institutionally unfriendly environment, may have some lessons to teach (Manzini 1996). After all, it is precisely because of the difficulties faced by "simply" transplanting successful institutional arrangements in different contexts that the whole field of policy learning studies has emerged. From our point of view, the Italian paradoxes may help illuminate the factors which are crucial to realize positive policy outcomes when everything seems to conspire against them.

In our study we have used both primary and secondary sources (see the Appendix for a list of interviewed actors and contacts), and adopted both a qualitative and quantitative approach. Whereas we try to offer a comprehensive view of environmental S & T policy-making, we also try to focus on specific details, partially giving up some generalization potential in order to reach a more in-depth understanding.

2. Backward institutions ...

Italy is among the seven most industrialized countries. In half a century, recovering from the Second World War, the "Italian miracle" has been realized thanks to a particular mix of big public companies in crucial sectors (chemicals, iron and steel industry, etc.) and countless small private enterprises. While still having a fairly consistent agricultural sector, the economy has gone through the traditional stages of industrialization, with the service sector gradually taking the lead position. This is to say that the country has experienced the typical pressures upon the environment which are connected to the development of the economy. Moreover, its geography, and the demographic burden upon a comparatively limited surface and the high levels of urbanization, have made it impossible to "hide" the consequences of this development for very long. Hence, environmental problems in Italy emerged at roughly the same time as in the other highly industrialized countries, and have been more or less similar in extent and complexity.

In spite of that, "Italy seems to arrive with a systematic delay at all the relevant steps in the development of an environmental policy, (...and) *ceteris paribus* with fairly different outcomes" (Lewanski 1997: 8). In that sense, and compared to what has been accomplished by comparable countries in terms of environmental responsibilities and tasks, "Italy has long represented an exception; it still is peculiar in many respects" (Signorino 1996: 113).

Let's look at the crucial phase in which most countries first developed the institutions responsible for tackling environmental problems, that is in the first half of the 1970s. Following the first awareness of air pollution dangers, something happened even in Italy: the Parliament established a provisional committee for ecological problems; the Minister for scientific research promoted the publication of the first report on the state of the environment; in 1973 there was even established an ad hoc department (without portfolio). But none of these steps succeeded in planting solid roots in the institutional framework, since they were all eliminated (in 1974) after one of the frequent governmental crises. After that, it took another thirteen years to reach the same point again.

Whereas the establishment of specific environmental departments is not a sufficient condition for effective policy-making (Dente 1995), and while it can be argued that specialization may sometimes be counterproductive (Knoepfel 1995), it is beyond any doubt that a few institutional ingredients have to be in place before considering their reform. In this perspective, the elements deemed essential in order to adopt consistent policy choices still represent a good indicator of the degree of political attention given to the environmental question, and of the potential for a sensible handling of the problems.

These elements should at least include:

- the introduction of an autonomous Ministry or Department;
- the establishment of an Agency for the environment;
- the adoption of a comprehensive legal framework for the different sub-fields;
- the regular publication of a national report on the state of the environment with data collected by a public agency.

According to the reconstruction made by Janicke (1991), most industrialized countries completed at least three of these four institutional steps during the 1970s, while Italy started to undertake the same trajectory only fifteen years later.

The first Minister for the environment was appointed only in 1985, and the department still awaits the completion of its staff endowment; its staff now mainly consists of non-technical officials temporarily appointed by other institutions (Salvia 1989). Meanwhile the (already poor) environmental budget has been often cut off, due to the general restrictions on public spending, and even because of the ineffectiveness of the ministry itself in spending its own resources.

The norm which established the National Agency for the Environment and provided the framework for enacting its regional counterparts has been approved in Parliament in 1994, partly as an answer to a controversial referendum which abolished the environmental competences of the NHS, and partly due to the international incentive represented by the European Agency. As a matter of fact, the national agency started to be operative only in April 1997, and just seven (out of twenty-one) territorial units have been legally set up, while even fewer are active (Frey 1996; Freddi 1997).

The amount of "environmental" laws issued both at the national and at the regional level is rather impressive. Unfortunately it is more the effect of a somehow confused and fragmented approach to problems than a measure of green concern on behalf of the political system. In fact, we had to wait until 1989 for the first comprehensive plan for the environment, while the most important measures adopted in each sub-field – air, water, noise, waste, etc. – show a delay which ranges from ten to twenty years (Lewanski 1992).

Finally, as for the regular collection and publication of environmental data, neither the Ministry for the Environment (which has produced only two complete analyses since its establishment) nor the National Institute for Statistics - which has done slightly better, publishing four reports since the mid 1980s – can claim to furnish reliable and updated sources of information. As we shall see, environmental associations can boast a much better record on this point.

We need not go further in portraying the inadequacies of the institutional framework set up to cope with environmental problems. It is nonetheless important to note that the Ministry for the Environment, in spite of its being poorly equipped in terms of internal technical competence, has neither been able to force the realization of specialized external structures on environmental science and technology[1], nor to establish regular relationships with more research oriented ministries[2], nor to steadily reorient the attitude of already existing technical bodies[3] towards more environmentally friendly activities.

The last clue needed to complete this introductory institutional overview has to do with the actions taken by the Italian government to implement its engagement towards a sustainable development, after subscribing to the Rio agreements in 1992. Before the end of 1993, each government had to submit to the UN its national plan for implementing the aims of Agenda 21. Strangely enough, the Interministerial Committee for Economic Planning (CIPE) succeeded in presenting a 150 pages document just before the deadline[4]. In that document, it is honestly stated that "the countries which, in their more or less recent past, have already devised programmes or strategies for the environment, are now facilitated in setting up their plans for the implementation of Agenda 21. Italy does not belong to this category".[5]

Unfortunately, but not surprisingly, this assumption of responsibility has not opened a new anticipative era. The document itself was explicitly addressed more to respond to national emergencies (p. 13) than to introduce a new approach in environmental policy-making (#10). Moreover, it received a cold-blooded welcome both from environmentalists and from the entrepreneurs. The former skeptically underlined how many parts contained commitments which were too vague in terms of the actions to be taken and instruments to be adopted. The latter, on the contrary, contested some restrictive provisions regarding energy consumption, waste production and industrial emissions on the basis that Italy was starting from a comparatively less consumption-oriented economy in comparison to other industrial nations, and further regulations would have a negative effect on competitiveness in the global market.

Furthermore, as for the institutional apparatus conceived by the plan itself, the projected committee for the coordination of Agenda 21 actions, which should have gathered information on the different initiatives and prepare an annual report on the accomplishments of the goals included in the plan, was never properly established. Local actions proceeded autonomously, with no central coordination or incentive. As we shall later see, we have to wait until the end of 1996, with the new Green Minister for the Environment appointing a consultative committee to see the process getting off to a (second) start.

3. ... and positive experiences

One of the most attentive observers of Italian environmental policy-making, using a computer metaphor, defined this policy area as still "un-formatted" (Lewanski 1990)[6]: there are no clear boundaries between competences, no proper limits signalling the trespassing of policy domains, no fixed rules of the game, no institutional legacy which constrains the actors in pre-defined roles, no monopoly over particular environmental cultures or assumptions. Far from being considered as a fruitful potential for flexible relationships and crossfertilization, this situation has often caused chaotic situations, communicative blocks, stop-and-go commitments, unreliable alliances, and so on. Moreover, the lack of institutionalization and of standard operating procedures has mainly brought personal relationships on the forefront. But environmentally-friendly Ministers can be substituted by more indifferent ones (especially where governmental turn-over is so frequent, as its is in Italy); eco-sympathetic entrepreneurs can't stand for a whole economic sector;

attentive local governments have no spill over effect if they rely entirely upon personal capabilities and preferences. If trust and information exchanges between policy actors have to be bargained on an *ad hoc* basis, because of the lack of an institutional framework which ensures inter-temporal committments and certainty, the range of possible agreements will be possibly broader (not being committed to coherent policy legacies) but with no guarantee as to the actual stability of the outcome.

In this context it may seem amazing that something positive happened at all. Without any claim to be exhaustive, we can identify mainly three classes of positive experiences:

- the first encompasses different kinds of agreement between social actors, but which do not include the central public authorities, in order to fulfill some environmental aim;
- the second class relates to the opening of public decision-making to the citizens' scrutiny and participation;
- and, under the third fall the unexpectedly brilliant results achieved by the public authorities on some environmental issues and the innovative solutions eventually introduced.

We will briefly give some examples for each class in order to call attention to the existence of numerous scattered exceptions to the general rule[7]. To the first class belong, for instance, all the experiences made by private and public firms to introduce some kind of eco-auditing and/or eco-budgeting. We still have to use the expression "some kind", because the Italian government has just identified the authorities in charge for the implementation of the EU Regulation 1836/93 regarding Emas, but they aren't yet operative (Peccolo 1997).[8] In spite of the legislative void, many firms of different size have decided to organize themselves with some academic body or think tank in order to anticipate the implementation of the EU regulation. Some have organized a "Club of Firms for Eco-Efficiency" supported by the Politecnico of Milano, a well-known university of engineering. Others have participated, together with environmentalists and civil servants, at the "Forum on environmental reports" organized by the Fondazione Mattei – an autonomous think tank mainly sponsored by Eni, a major public company. In 1994, a project submitted by the main entrepreneurial association of the chemical sector, Federchimica, together with a research institute of the Bocconi university, has been chosen as one of the few EU pilot program in eco-auditing, to explore its application to small and medium entreprises.

Other cooperative initiatives not sponsored by public authorities include new concerted projects between environmentalists and trade unions, both at the national and local level, to extend the use of alternative energy sources, and to support the introduction of green productions which should be able to create new work opportunities; the so-called "Eco-sportello" project, an agreement between *Legambiente* (one of the most active environmental organizations), the association of the municipalities and two organizations which represent the local environmental services to promote the adoption of new technologies in the field of waste management, and activities such as collecting data, informing citizens, spreading information; some joint undertakings between environmental and entrepreneurial associations to by-pass the political or bureaucratic inertia, especially at the local level (#4); the joint sponsorship of specific events and campaigns, though often for mere symbolic reasons from behalf of the involved companies.

The few cases of Local Agenda 21 certainly pertain to the second class of successful occurences – the one directly concerned with public participation in the definition of the contents of policy-making. A typical example of positive experience is represented by the Eco-label received at the 1996 Lisbon Conference on sustainable cities by the municipality of Bologna for its commitment toward the realization of the aims of the Aalborg Charter (Berrini 1996). The same municipality participated in the Urban CO_2 Reduction Project coordinated by the International Council of Local Environment Initiatives. Though still a minority, there are quite a few other local projects regarding energy, transport, air pollution, waste management, etc. which make provisions for the participation of the public. Most of the time, they are aimed at increasing the democracy and the transparency of policy-making processes (e.g. the EIA); environmental groups or locally organized citizens are frequently entitled to supervise the accurate implementation of specific projects; less often they are consulted about the aims of the policies which should be taken, or included in the discussion regarding the technical aspects of the policy.

At a completely different level, but still regarding the participation of the public, we should remember the use of popular referenda. They have been organized on many different topics, and with different aims. There have been local and national referenda; abrogative and consultative; accepted and rejected by the High Court; on nuclear energy, environmental competencies, hunting, pesticides, traffic, localization of firms, etc. In spite of the low levels of mobilization among the general public normally assured by these kind of issues, this instrument is usually very effective for the environmentalist interests.

Finally, the third class of positive histories includes all the examples of successful policy innovation. These experiences range from the successful waste management of the municipality of Milan, which almost reached German-levels of recycling in a short time span and starting from an emergency situation, to the implementation of automatic systems for controlling private mobility and traffic (Bologna), to the urban improvements following bottom-up pressures in cities like Neaples, Rome and Genoa, to experiments in "time-regulations" aim at reducing air-pollution, to generally attentive local administration whose green concern and positive outcomes have been yearly monitored by a comprehensive research on urban ecological systems realized by an environmental think tank.[9]

But positive experiences of this sort are not to be found only in the public domain. In the private sector, for example, there are many cases of introduction of industrial cleaner technologies. These have been both analysed in the literature through a case study methodology (e.g. Dente and Ranci 1992; Malaman and Paba 1993; Gerelli 1994), and tentatively mapped in a seminal paper sponsored by the already quoted Fondazione ENI-Mattei (Malaman 1996a). These technological innovations implemented by firms in different sectors and of different size, are to be found in all parts of the country (though they are mainly concentrated in the North), and equally address the most common environmental problems.[10]

We are well aware of the fact that the few examples here reported can't substantially affect the average backward image portrayed before. Still they break down the homogeneity implied by that framework, suggesting the existence of a much more variegated situation. How to account for that variety is a matter of concern for the following paragraphs.

4. Policy domains

In the PESTO project, S & T environmental policy-making is conceptualized as an arena of interaction between four domains – economic, bureaucratic, academic, and civic.[11] Each domain is animated by different actors, characterized by different ideals and, possibly, by different attitudes to the general social functions of science and technology innovations (Elzinga and Jamison 1995). The differences in the assumptions and kind of arguments perceived as legitimate in the different domains should lead to different preferences in the choice of policy instruments (Liberatore 1995), although these may be more easily subjected to bargaining (Sabatier and Jenkins-Smith 1993).

4.1 Bureaucratic domain

The interface between the "public" and the "policy" sees policy-takers and policy-makers on different, though communicating, fronts. Stakeholders are mainly citizens and entrepreneurs, mostly interested in the defense of public goods and private incomes. On the policy-delivery side, politicians and bureaucrats share some common concern, though they normally respond to different *stimuli*, such as electoral consensus and formal accuracy.

It has been pointed out that there are quite a few inconsistencies between these aims, which mainly reflect a deeper contradiction between sustainable development and the traditional models of liberal democracies (Dente 1994). Pointing to possible limits to growth has serious consequences in terms of margins enjoyed by policy-makers in formulating distributive and redistributive policies, that is, in the way they normally secure their consensus. The protection of public and positional goods (Hirsch 1976), that is of commodities which are exposed to indiscriminate consumption or congestion, calls for regulatory powers that are becoming "outmoded" in a market-oriented style of intervention. Sustainable development concerns even the well-being of future generations, the ones on which decision-makers have traditionally deferred the costs of their present policies. The globalization of environmental problems requires an internationalization of its governance, which tends to be exposed to national defections. Traditional democratic forms of representation are mostly ineffective in reducing internal and external costs of decision-making (Sartori 1987): opening up the policy-making arena to the contribution of an increased number of participants may still be assumed to reduce the amount of external risks (Beato 1994), but the legitimacy for closing it down can't simply rests upon the classic territorial mandates (Osti 1994), since it has to take into account scientific arguments, inter-generational stakes, collective action shortcomings, etc.

Politicians seem less worried by the growing concern for public participation in environmental matters than by short-term affairs. With the obvious exception of the green party, electoral platforms pay a symbolic lip service to environmental matters, and there is almost no trace of interest in "sustainable development" in the parliamentary records: in the last two decades, MPs have submitted only four bills addressing this topic, and the environmental parliamentary committee has never issued specific public hearings, not even in the event of the Rio conference. Things may be different at the local level, especially at the regional one, where many environmental competencies have been transferred, but, apart from scattered examples, mostly derived from personal acquaintance (#3, #6, Diani 1996), politicians seem to believe that they have more to lose than to gain from increased public participation and transparency in policy-making processes.

Generally speaking, the civil service is even more worried about this possibility (Donati 1994). Though our respondents sometimes recognize the potential for consensual resolution of problems implicit in opening up the policy-making arena to citizens, consumers or environmental groups (#5), still they fear the possible chaotic and inconclusive outcomes of that participatory approach. Many remark that a more inclusive style is theoretically sound, but doesn't work in practice; at least, not in Italy.[12] This is all the more true for ambitious environmental plans like Agenda 21 and sustainable development (#8, #10, #12):

> "The truth is that, in Italy, tasks and responsibilities have become more complex: Regions have been established, then the local councils, associations have been founded, the unions joined in, there are a lot of parties, of departments,... the truth is that, now, there are too many people who operate in the environmental field, and we hamper each other ... and so the environment stares at us and laughs, and takes its own revenge" (#11).

> "The number of potential actors in a policy for sustainable development is so high, that we have somehow to split the job ... the risk is to establish committees made by hundreds of people that don't achieve anything. ... The National council for the environment itself is too wide: there are 20 regions, 15 environmental groups, etc. it is too wide, they are too many. ... To have consultation procedures is necessary, but then the public service should have the capacity to proceed as it believes to be necessary" (#9).

> "I can't see how one could further extend citizens' power in this country ... any local council can call referenda to stop any project, one can't go any further" (reported in Diani 1996).

Not surprisingly, apart from the environmental department, green groups are often considered the major challengers. In spite of the reduced confrontational style (Diani 1996), and of the fact that conflicts mainly originate on single issues at the local level (#1), they are often still portrayed as stubborn opponents by central civil servants:

> "environmentalists are always against" (#8);

> "... there is a kind of exasperated environmental fundamentalism" (#10)

> "... at the end, even the environmental world has become a political movement, a movement of power ..." (#11).

Whereas the suspicious attitude towards the civic domain is almost palpable, dispositions towards the scientific and economic domain are multi-faceted. Experts are deemed necessary, because technical competencies have always been underdeveloped in comparison to juridical and administrative ones (#9); still, there is sometimes some skepticism regarding their external contributions to policy-making:

> "The world of (environmental) research isn't yet sufficiently organized" (#12).

> "(at the time of the Chernobyl accident), in the emergency committee we were four ignorants, because the world of science had abandoned us" (#11).

Economic actors – private companies, entrepreneurial associations – are normally perceived as potentially relevant partners, probably because they are thought to be right in pointing at the overall inefficiencies of the public administration (#9). Nonetheless, there isn't any institutionalized forum which includes economic actors. Even the recently established working group for the implementation of Agenda 21 does not encompass socio-economic actors: unions, entrepreneurial associations, etc.; this, in spite of the fact that its coordinator suggested to include them (#4):

> "the dialogue with the industrial counterpart should be much more strengthened ... though it is a complex operation, which shouldn't be only one way" (#9).

At the same type, according to Donati (1994: 188), the stereotyped generalization "business=environmental exploitation" is sometimes perceived to be true: "some public administrators describe an unusually high resistance, among Italian entrepreneurs, to technological innovation, and in particular to innovation on environmental protection measures".

4.2 Economic domain

The economic domain is inhabited by very different actors: unions and entrepreneurial associations, firms acting in different sectors, hazardous production and eco-business, environmentally-friendly entrepreneurs and unresponsive polluters. The variety is somehow an element which is intrinsic to the market.

This is all the more true for the Italian economic domain, with its north-south cleavage, its complex public-private structure, its big-small entreprises antagonism, its inter-generational differences (#7, #2).

Hence, it is difficult to look for a common perspective inside this variegated domain; it is even questionable whether the size of the firm favours or inhibits the adoption of environmental S & T innovations (#7, Donati 1994, Malaman 1996a). Nevertheless, it is generally acknowledged that the industrial sector has nowadays a less confrontational approach towards the green movement (Cavallo 1997).

> "With the ecologist world there is communication ... on many issues we have to find a common ground on which to discuss, because, otherwise it is impossible to find a solution"

> "It is worthwhile because they have brains, heads, competencies, positions. It is worthwhile because the environmental issue concern directly the relationship between the firm and the society, and they are the society ..." (both quoted in Donati 1994: 122).

Environmental associations are sometimes regarded as more reliable partners than the public administration itself (#4): discussions are said to be related to substantial problems, to different technical evaluations, and not to some formal bureaucratic aspect in fulfilling complex norms. "With a few remarkable exceptions, public officers tend to be perceived as passive actors in these policy networks, where most initiatives come either from environmentalists or business representatives" (Diani 1996: 9).

Obviously, private interests aren't excluded from the informal consultations which normally precede any relevant policy decision, but entrepreneurial associations would probably prefer more institutionalized relationships:

> "I would say that I am satisfied (with the degree of involvement of the association in public policy-making) but I believe we should do more, in the sense that it should be possible to establish more formal arenas in which ... we could explain our positions better, and we could make our observations and proposals" (#7).

In the last few years, thanks to the scientific turn of a part of the environmental movement and their improved relations with the trade unions, as well as the steering influence exercised by the European community, private entrepreneurs have faced growing difficulties in lobbying effectively. At the same time, they have often suffered from bad reputation, needing to campaign actively to improve their public image (e.g. the chemical industry which is particularly active on that front). Both these reasons could account for the request of more institutionalized relations, since the least desired outcome is the unrestricted participation of citizen in environmental decisions:

> "First of all, I do not believe in direct democracy; denouncing a problem and then let citizens express themselves and decide by simply pressing a button: I think this one is the perfect solution for preventing the country to improve" (#7).

Whereas in the interface with the policy-making arena, entrepreneurs seem more willing to adopt an anticipative and consensual style, they often appear passive and isolated as far as the introduction of "green innovations" in the production cycle is concerned. An extended survey of the implementation of cleaner technologies has demonstrated that the main reason for environmental innovation is still the influence exercised by regulations, and that most of the firms (excluding the big multi-national corporations) adopt defensive strategies with regard to standards, instead of pro-active ones. Moreover, "cooperation agreements between business, universities and public and private laboratories are the exception rather than the rule in the development of environmentally friendly technologies" (Malaman 1996a: 22). Apparently, to use a sort of aphorism, whereas there seems to be a social role for economic actors, there isn't any economic role for social actors.

4.3 Academic domain

The academic domain is the realm of science and expertise: purely endogeneous forces steer its progress and the judgement made are impartial. Probably we wouldn't need Lakatos or Feyerabend to challenge this over-simplistic perspective from an epistemological angle; surely we don't need them to challenge it empirically. The literature on S & T innovations helps us clearing off our understanding from the naive equation: "scientific contribution to decision-making" equals "search for the rational solution".

Still, one of the essential resources for legitimizing the choice of specific policy instruments is "scientific evidence", or at least the claim to portray some empirical evidence as the true scientific one. The famous quotation by Schattschneider (1960) that "problem definition is the supreme instrument of power" perfectly applies to this policy area. Through the last two decades there has been a growing awareness that scientific knowledge wasn't only a technical element in instrumental choices, but a political resource in policy-making processes. At least at this level, environmental expertise has become a business.[13]

At the end of the '80s, actors from the bureaucratic, economic and civic domain, besides trying to frame their preferred solutions under the appealing label of science and rationality, have slowly built their own environmental competencies. The Ministry for the environment has promoted a plan for S & T research, involving ENEA (the national institute for energy), CNR (the national research committee) and the ISS (the high institute for health studies). Many regional governments have activated (formally) private research centers. The major entrepreneurial association – *Confindustria* – has sponsored an authoritative journal (*Impresa Ambiente*), and a new institute for environmental research (IPA – *Istituto per l'ambiente*). An already active group of green researchers, mostly from *Legambiente* (see further), has organized itself for offering environmental advise mostly to local governments, and for monitoring the state of the environment (*Istituto di Ricerche Ambiente Italia*).

It is hard to outline the upsurge of initiatives which involved so many research centres, mostly with some private consultancy by academic people. We can obtain a sort of "map" of this scientific universe by reviewing the database on environmental research projects compiled in 1994 by the *Fondazione Lombardia per l'Ambiente* – a foundation established with the compensations obtained after the Seveso accident and mainly working for the regional government.[14] Checking the main partners of each research project, we can give a first glance at the research production in the environmental field, most of which is done on S & T topics.

Actor	single	multi	Actor	single	multi
University	127	163	Public research inst.	47	37
Local government	2	60	Private research inst.	40	95
Public Administration	49	61	Foreign research inst.	7	50
C.N.R.	491	23	Hospital and NHS	3	40
I.S.S.	11	114	Others	–	9

Tab. 4.1
Number of environmental research projects (alone and in partnership).[15]
Source: Own elaboration on data of the Fondazione Lombardia per l'Ambiente.

Scanning the database we identified 10 distinct classes of actors, and we classified their research activities according to whether they had been conducted alone or with other partners: for example, while the National Research Committe (CNR) normally "works for itself", the Institute for Health

Studies (ISS) develops several partnerships. In this last case, it could be interesting to investigate which scientific relationships tend to emerge inside this academic domain. Figure 4.1 portrays exactly the most significant of these relationships, while the positions occupied within the scattergram – whether in middle or at the margins – can be intuitively associated with the central or peripheral role played in these scientific networks.[16]

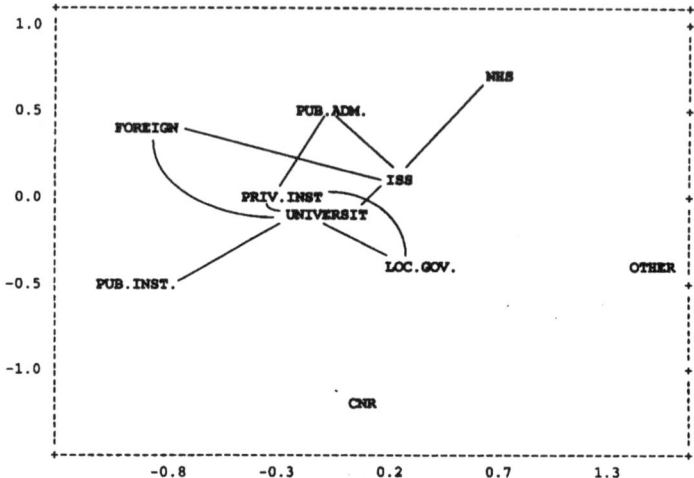

Fig. 4.1 A tentative map of the academic policy domain.

Universities certainly represent the heart of this domain, fostering stable research connections with many other actors; local governments, public and private institutions, foreign think tanks and the ISS are the traditional partners. But they don't represent an homogeneous epistemic community (Haas 1989). At a closer look, we would argue that university researchers are mainly co-opted by other subjects in order to reassure them on the methodology of some applied research, and to legitimize its results: scientific arguments are used to advocate positions more than to falsificate hypotheses.

The ISS seems to be the other core actor of the system, probably as the effect of the health legacy of the Italian environmental policy. However, its role is probably opposite to that of the Universities. The ISS sponsors research more than doing it: is on the demand side more than on the supply side of the environmental research market. The same distinction applies for private and foreign research institute – which are similar to universities – and for local governments and public bodies – which correspond to the ISS situation. Other actors seem more peripheral, either for being less involved in applied research (e.g. CNR), or for maintaining selective and issue specific links with a few actors (e.g. Hospital and local NHS, mainly connected to the ISS). Finally, it is interesting to note that public and private research institutes do not make a great difference for local governments or public administrations: they are both acceptable partners, with a slight preference for the private associates.

Beyond the political exploitation of empirical evidence, which is a constant in policy-making processes (Stone 1988), scientific institutions suffer a lack of direct contact with lay people. Their "interface" to the general public, that is mass media, is normally tuned on a different language, and, in Italy, it traditionally assures a poor coverage of the more scientific aspects of environmental policies (Borrelli 1996). This may partially account for the doubting disposition exhibited by citizens towards the environmental progress assured by science.

4.4 Civic domain

In the civic domain we must mainly account for two different phenomena: the attitudes exhibited by the general public towards environmental problems – their wish to be included in decision-making, their confidence in scientific solutions, their (lack of) trust in institutions, their mobilization potential – and the role played by social movement organizations, mainly environmental associations (Donati 1994).

As everywhere, in Italy the NIMBY syndrome is quite common. Local groups of citizens activate themselves in order to stop the realization of power plants, installations for waste management, hazardous industries, motorways, etc. Since they are seldom consulted before, the mobilization mainly regards the implementation of these projects, sometimes at an irreversible stage. The Italian geography aggravates this syndrome: actually, there are only a few areas which are not urbanized, or environmentally protected (mountains and coasts), or agriculturally employed, or tourist oriented.

Unfortunately, even when enlightened institutions decide to involve citizens in earlier stages of the process, they face a population which mostly distrusts them. Previous experiences have eroded the necessary "trust capital" (Ascari et al. 1992), and even projects at the highest standards of protection may fail to gain the citizens' approval. Lay people have long learned to doubt the promises of politicians, but in the near past they have even had good reasons to question scientific evidences. A recent comparative survey on the attitudes of citizens facing the environment shows this generalized attitude quite clearly (see tab. 4.2).

	Italy	Germany	Great Britain	Netherlands	Norway
Institutions	70.2 %	1.7 %	38.0 %	2.1 %	n.a.
Science	32.2 %	27.1 %	26.4 %	14.0 %	11.2 %
Economic growth	56.2 %	51.0 %	25.8 %	30.2 %	19.1 %

Tab. 4.2 Indicators of mistrus t.[17]
Source: ISSP 1993.

There is no field, that is no legitimizing assumption, in which Italian respondents do not exhibit their total lack of confidence. The other countries present different mix of attitudes, following citizens' perceptions of what they consider the most promising approach to environmental problems: in Germany they seem to rely upon governmental regulations, whereas in Great Britain they seem to trust more the market, and in Netherlands a mix of the two ; in Norway, the reported data picture a situation of greater confidence both in scientific and economic progress. Anyway, none of these countries comes closer to the levels of suspicion displayed by Italy, not even in their least trustworth domain. It is not a surprise, then, that whenever citizens are consulted *ex-ante*, e.g. through referendum, the consultation turns out to be an environmentalist victory. Yet, the feedback is that (even green-sympathetic) administrations are more and more reluctant to include public participation in the earliest phases of decision-making processes, because they fear to be stalled in every project, thus increasing the level of skepticism and starting the vicious circle again.

> "Here, this kind of arrangements (consultation procedures and public hearings) are not as common, probably because of the weaknesses of the civil service itself: when a public administration decides to start exchanges and concertations with citizens, it isn't any more able to close them; il lacks the capacity to close them, it has no indipendence, no autonomy, no technical strength to do that, not because it is impossible to do it" (#9).

At a non-local level, this attitude may transform itself in a general feeling of ineffectiveness, thus emptying the mobilization potential for broader political actions guided, for instance, by environmental groups. In fact, the quoted ISSP survey portrays a population apparently willing to use every possible instrument to protect the environment – Italians seem to be comparatively readier to pay more taxes (!), higher prices and accept more regulations – but which, at the same time, seems to be unconfident towards any form of activism – they don't support environmental groups, don't donate money for environmental campaigns and they don't sign petitions.

In this framework, even under the former proportional electoral system, green parties attracted more consensus than votes and, on top of that, they were even considered with some suspicion by grass-roots groups (Poggio 1996). The environmental movement itself was split into several quite distinct "currents" (Biorcio and Lodi 1988; Diani 1995), and, partially for this reason, had not the same institutionalized chances of pressure-making it had elsewhere (e.g. Jordan and Maloney 1996).

Its major groups were the Italian branches of WWF (mostly with a conservationist approach), that of Friends of the Earth[18] and of Greenpeace (with their more radical single-issue actions), "Italia Nostra" (modeled upon the British National Trust, and, together with the "Fondo per l'ambiente italiano", largely with the same aims), and Legambiente.[19] This last association, which largely picked up the legacies of preceding new-left wing and anti-nuclear movements, has gradually become the most active group in the scientific environmentalism, attracting a growing number of associates and working at close contact with sympathetic members of the academic community. Its scientific committee includes more than a hundred professionals with all the needed expertise – some working for the association, while mostly consulted (free) on a regular basis (#2).

Its scientific approach has brought this association to develop closer contacts with the other domains, sometimes due to personal ties or overlappings (#3, #4), sometimes following the development of common projects, sometimes as valid substitutes to formally responsible institutions, but mostly for not being considered any more as maximalist anti-development advocates (Diani 1996). This attitude towards a technical understanding of problems, coupled with the capillarity of its structure, permitted Legambiente to produce yearly reports on the state of the environment which, starting from 1989, quickly became more authoritative and updated than the ones irregularly published by the Environmental department. The already quoted Istituto di Ricerche Ambiente Italia, though independent, draws its founding members mostly from the steering committee of Legambiente (#4), and it can be practically considered as its own think tank.

Besides gaining the regard of actors normally considered antagonists – e.g. entrepreneurial association – the "success" of the "strategy" of Legambiente had two major spill-overs. On one side, the recognition of environmentalists as legitimate subjects in institutional working groups or as independent national experts in internal contexts, acting not only as representatives of stakeholders but as bearer of technical solutions (#1). Secondly, the spreading of this approach even to more traditional green associations (#2, see e.g. Signorino 1996; WWF Italia 1996).

5. Conclusions

The Italian environmental S & T policy-making is marked by a singular puzzle. A poorly defined institutional framework has not inhibited the emergence of a few successful experiences, These are all the more interesting because they may help in detecting non-formal factors which permit positive outcomes within adverse contexts.

The evidence collected in our analysis can be summarized as follows:

- Rules – intended as a coherent set of authoritative institutions aimed at regulating the environmental intervention – are neither sufficient nor (sometimes) necessary elements for successful policy histories. We have argued elsewhere that similar hard institutional arrangements may result in very different outcomes (Giuliani 1998), this being an argument against short-sighted imports of formal structures. Here we have noticed how micro-local positive experiences can be achieved in spite of a macro-institutional backward context.

- A constant in these positive experiences seems to be the presence of consensual pattern of relationships between policy-actors or, at least, a greater propensity to maintain a larger web of links connecting formally responsible decision-makers to policy-takers and the interested public. In this sense, the attitudes exhibited by policy-makers themselves, that is their cognitive maps, how they define problems and solutions, who they recognize as legitimate partners in decision-making processes, turn to be important elements of these processes.

- These links "functionally" substitute the institutional frame in assuring the required stability for the actors' interactions: they ensure the possibility of mutual adjustment through the development of iterated relationships and, while extending the time horizon of each "game", may promote more trustful interchanges. At the same time, links (more than institutions) can perform the role of consensus-keeping, internalizing the costs of potential opposition. Positive-sum games aren't always attainable, and sometimes top-down decisions are the only way out from stalemates; nevertheless, a constantly secluded approach to decision-making normally inhibits the achievement of higher levels of satisfaction.

- Policy entrepreneurs may help in bridging the resistance which normally oppose actors belonging to different domains, and which traditionally inhibits more productive relationships. They can perform different types of action: first of all, they promote new policy-paradigms, around which rearranging prior solutions; secondly, they ensure the communication between actors which usually talk in "different languages"; thirdly, they guarantee the "fairness" and "reciprocity" of the exchanges. Entrepreneurship, however, has to routinize itself in order to survive contingent turnovers (e.g. the change of Minister), and to leave a constant footprint in policy processes.

- This has not frequently happened in the Italian environmental policy-making. For this reason, whereas on single public interventions, informal and contingent nets can find adequate solutions quite effectively, this turns out to be more difficult on strategic policies like Agenda 21. In these cases, a cooperative and professional bureaucracy (and hence, the formation of civil servants) becomes all the more important. Broad-aimed policies probably require entrepreneurial attitudes in linking disparate competencies and viewpoints, but last only if supported by a cooperative administrative apparatus.

Appendix
Interviews and contacts:
1. Senior civil servant National Agency for the Environment
2. Head of the Scientific committee of Legambiente
3. Deputy head of Legambiente Lazio, elected of the Green party at the Municipality of Rome and legislative consultant of the Green parliamentary group
4. President of the Istituto di Ricerca Ambiente Italia, member of the DGXII expert group on the Urban development, and coordinator of the National working group on sustainable cities of the Environmental department
5. Senior civil servant, Environmental department, Regione Lombardia
6. Former "assessore" for the environment Regione Lombardia
7. Former member of presidential board of Confindustria
8. Senior civil servant, Transport department
9. Senior civil servant. Environmental department
10. Senior civil servant, Industry department
11. Senior civil servant, Agriculture department
12. Senior researcher ENEA

Consultant responsible for the Agenda 21 of Modena
Senior civil servant responsible for the Agenda 21 of Bologna
Senior civil servant responsible for the Agenda 21 of Genova
Coordinator of the committee for the Agenda 21 of Milano
Leader of environmental association and member of the committee for the Agenda 21 of Milano
Senior researcher of the National Research Committee

Notes

[1] The scientific committee which should have given its advise to the Ministry has been abolished, and the National Council for the Environment, composed by experts, environmentalists, representative of local governments and some public agencies, has never been really effective (besides, its meeting have become more and more rarefied).

[2] Actually, in 1990, the Minister for the Environment has promoted, together with the Ministry for the University and for Scientific Research, a "National Plan for the environmental science and technology research", with an initial budget of around $ 150 millions, and establishing the Scientific Committee for the High Consultancy in Environmental Research; but the whole project has not left any remarkable sign on the governmental environmental policy-making.

[3] The ENEA (which mainly worked on energy, and, eventually, formed the core structure of the Environmental Agency), the Higher Institute for Public Health, and, partially, the National Research Committee, were the most plausible candidates in this regard, but probably their respective legacies prevented them from playing that role in an active manner.

[4] The exact date is 28 December 1993, although it was published in the Official Journal only two months later.

[5] CIPE, "Piano nazionale per lo sviluppo sostenibile in attuazione dell'Agenda XXI", Official Journal, n. 47, 1994: 12.

[6] "The policy arena is not yet considered stable by the same involved actors: new players enter the field with an active role in it; standard operating procedures have not yet been established; there isn't a clear policy agenda; new problems, goals and solutions are constantly emerging" (Ibid.: 309).

[7] The municipality of Ferrara, together with Legambiente, has organized an extensive database of positive environmental programmes which is called OPERA ("Osservatorio Permanente sull'Ecosistema Urbano") and which can be easily consulted on Internet at http ://www.comune.fe.it/ambiente/index.htm

[8] In Europe, Italy is in third position, after the United Kingdom and Netherland, as for BS-7750 and ISO-14001 certifications, but is last regarding the European Eco Management and Audit Scheme (Cavallo 1997; Legambiente 1997).

9 This research, realized each year since 1994 by the "Istituto di Ricerche Ambiente Italia" following the idea of territorial eco-reporting or budgeting, covers more than 100 cities (all the Italian provinces) analysed by means of 20 different environmental indicators. Its results have been published in Legambiente (1995; 1996a; 1996b).

10 The introduction of cleaner technologies in the sample of 192 examined innovation processes were aimed at the reduction of air pollution in 34% of the cases, of waste in 33%, of water pollution in 29%, and at the reduction of noise in 4% of the occurences.

11 Much in the same mood, Malaman (1996: 12) suggests that "technological change for the purpose of environmental conservation has the characteristics of an endogenous process, dependent upon economic incentives, technological opportunities, and socio-institutional relationships".

12 They sometimes end up saying something like "... if only Italians decided once and for all to behave themselves properly, we wouldn't have problems any more" (#8).

13 The percentage of public spending in the field of environmental research has grown three times during the '80s; still its absolute amount is not comparable to that of countries like Germany. Environmental research sponsored by privates sum up to the same value, that is around 150 millions $ in 1991 (Malaman and Paba 1993).

14 This database includes brief summaries of more than 800 environmental research project mainly located in the Lombardy area. The geographical limitation shouldn't affect analysis, since most of the research institutes and think tanks working on the environment are actually located in this region. Moreover, a study on biotechnology has demonstrated that most of the research conducted in Italy on that issue is realized in Lombardia (Orsenigo 1991).

15 CNR stands for the National Research Committee, while ISS means High Institute for Health Studies. Under the label Public administration we have classified all the research projects conducted by governmental departments or by public companies (such as Enel, Eni, etc.).

16 Positions and links relate to research networks, that is to projects conducted together by different partners, and do not necessarily correspond to the overall relevance of each actor (e.g. CNR is pictured at the periphery despite its 400 projects). The exact locations in figure 4.1 are actually the result of a MDS analysis based on the amount of projects carried on together by each couple of actors; the underlying dimensions could be a specialism-generalism continuum on the north-south axis, and a scientific-tecgnic one on the west-east axis. Relationships have been considered significant – and traced in the diagram - whenever they implied more than ten joint projects.

17 The table reports the percentages of people who "strongly agree" or "agree" regarding different statements. For science: "Overall, modern science does more

harm than good" (similar results are obtained asking if people believe in science too often ...); for economic growth: "Economic growth harms the environment". The question regarding institutions includes those respondents who think that "their country's effort to protect the environment" is a bit less or a lot less than what other countries do. Answers for Germany include only West Germany citizens.

[18] Actually a minor group near to the Radical party, mostly pressing for referendum initiatives and for the establishment of the environmental agency. Eventually, when the agency was approved, the leader of the environmental association became its president.

[19] There are obviously more environmentalist groups, mainly specialized on different issues: the number of associations recognized (and partially financed) by the Environmental Ministry has grown from 17 in 1989, to 27 in 1994.

References

Ambiente Italia (1996) *Agire localmente, pensando all'Europa*, Conference documents, Roma, 21 June.

Ascari, S. et al. (1992) *L'igiene urbana. Economia e politica ambientale*, Angeli.

Beato, F. (1994) 'La progettazione di istituzioni per la salvaguardia dell'ambiente: il dilemma tra efficacia e partecipazione pubblica', in L. Pellizzoni e D. Ungaro (eds.), *Decidere l'ambiente*, Angeli.

Berrini, M. (1994) 'Domestic Environmental Politics in a Comparative Perspective: The Italian Case', in O. Höll (ed.), *Environmental Cooperation in Europe. The Political Dimension*, Westview Press.

Berrini, M. (1996) 'La campagna europea Città Sostenibili e le iniziative italiane', *Caos*, n. 7.

Biorcio, R. and Lodi G. (1988) (eds.), *La sfida verde*, Padova, Liviana.

Borrelli, G. (1996) 'Metti il clima sul giornale', *Impresa ambiente*, n. 8.

Cavallo, M. (1997) (ed.), *La gestione ambientale nelle imprese*, Angeli.

Comitato Agenda 21, *Agenda 21 per Milano. Idee e proposte per Milano sostenibile*, Committee documents, Milano, January.

Dente, B. (1994) 'Sviluppo sostenibile e democrazia sono compatibili?', *Queste istituzioni*, 22, n. 97.

Dente, B. (1995) (ed.), *Environmental Policy in Search of New Instruments*, Kluwer.

Dente, B. and Ranci P. (1992) (eds.), *L'industria e l'ambiente*, Il Mulino.

Diani, M. (1995) *Green Networks*, Edinburgh, U.P.

Diani, M. (1996) *Environmental sustainability and institutional innovation in Italy*, IARD Research Report.
Donati, P. (1994) *Framing and Communicating Environmental Issues*, Research Project 42, European University Institute.
Elzinga, A. and Jamison, A. (1995) 'Changing Policy Agendas in Science and Technology', in S. Jasanoff et al. (eds.), *Handbook of Science and Technology Studies*, Sage.
Freddi, G. (1997) *Le agenzie per l'ambiente*, Angeli.
Frey, M. (1996) 'Il silenzio dell'Arpa', *Impresa Ambiente*, n. 4.
Gerelli, E. (1994) (ed.) *Applicazioni attuali e tendenziali delle tecnologie pulite: prime indicazioni e politiche per la loro diffusione*, Istituto per l'Ambiente.
Giuliani, M. (1998) 'Soft institutions for hard problems. Implementing air pollution policies in three Italian regions', in W. Grant, P. Knoepfel and A. Perl (eds.), Aldershot, Edward Elgar (forthcoming).
Haas, P. (1989) 'Do regimes matter? Epistemic communities and Mediterranean pollution controL', *International Organization*, vol. 43, n. 3.
Hirsch, F. (1976) *Social Limits to Growth*, Twentieth Century Fund.
IReR (1996) *Condizioni per uno sviluppo sostenibile in Lombardia*, Milano.
Jamison, A. (1996), 'The Shaping of the Global Environmental Agenda: The Role of Non-Governmental Organisations', in S. Lash, B. Szerszynski and B. Wynne (eds.), *Risk Environment and Modernity*, Sage.
Janicke, M. (1991) *Institutional and Other Framework Conditions for Environmental Policy Success. A Tentative Comparative Approach*, Environmental Policy Research Unit, Free University of Berlin.
Jordan, G. and Schubert, K. (1992) 'A Preliminary Ordering of Policy Networks Labels', *European Journal of Political Research*, vol. 21, n. 1: 95-123.
Jordan, G. and Maloney, W. (1996) 'How Bumble-bees Fly: Accounting for Public Interest Participation', *Political Studies*, vol. 44: 668-685.
Joy, L. (1996) 'Agenda 21 locale per migliorare la città', *Impresa ambiente*, n. 6.
Knoepfel, P. (1995) 'New Institutional Arrangements for the Next Generation of Environmental Policy Instruments: Intra- and Interpolicy-Cooperation', in Dente (1995).
Legambiente (1995), *Ambiente Italia 1995*, Edizioni Ambiente.
Legambiente (1996a), *Ambiente Italia 1996*, Edizioni Ambiente.
Legambiente (1996b), 'Ecosistema urbano 1996', *Caos*, n. 7.
Legambiente (1997a), *Ambiente Italia 1997*, Edizioni Ambiente.

Legambiente (1997b), *Cittadinanza, ambiente, sviluppo*, Conference documents, Roma, 89-9 April.
Lewanski, R. (1990) 'La politica ambientale', in B. Dente (ed.) *Le politiche pubbliche in Italia*, Il Mulino.
Lewanski, R. (1992) 'Il difficile avvio di una politica ambientale in Italia', in Dente and Ranci (1992).
Lewanski, R. (1997) *Governare l'ambiente*, Il Mulino.
Liberatore, A. (1995) 'Arguments, Assumptions, and the Choice of Policy Instruments', in Dente (1995).
Liberatore, A. and Lewanski, R. (1990) 'The Evolution of Italian Environmental Policy', *Environment*, vol. 32, n. 5.
Malaman, R. (1996a) *Technological Innovation for Sustainable Development: Generation and Diffusion of Industrial Cleaner Technologies*, Working Paper 66.96, Fondazione Eni Enrico Mattei.
Malaman, R. (1996b) 'Quali servizi idrici per Milano?', *Impresa ambiente*, n. 9.
Malaman, R. and Paba, S. (1993) (eds.) *L'industria verde*, Il Mulino.
Manzini, E. (1996) 'Azienda Italia per caso eco-virtuosa', *Impresa Ambiente*, n. 8.
March, J. and Olsen, J. (1989) *Rediscovering Institutions. The Organizational Basis of Politics*, The Free Press.
Orsenigo, L. (1991) *Archipelago Europe. Islands of Innovation. The Case of biotecnology in Italy*, Fast, FOP 243, EC.
Osti, G. (1994) 'Forme di legittimazione sociale del governo dell'ambiente' in L. Pellizzoni e D. Ungaro (eds.), *Decidere l'ambiente*, Milano, Angeli.
Peccolo, G. (1997) 'Gestione ambientale. Il sistema Emas nei Paesi europei', *Impresa Ambiente*, n. 4.
Poggio, A. (1996) *Ambientalismo*, Milano, Editrice Bibliografica.
Richardson, J. (1982) (ed.) *Policy Styles in Western Europe*, George Allen & Unwin.
Sabatier, P. and Jenkins-Smith, H. (1993) *Policy Change and Learning. An Advocacy Coalition Approach*, Westview Press.
Salvia, F. (1989), *Il Ministero dell'ambiente*, La Nuova Italia Scientifica.
Sartori, G. (1987) *Theory of Democracy Revisited*, Chatham House.
Schattschneider, E. (1960) *The Semisovereign People*, The Dryden Press.
Scott, J. (1991) *Social Network Analysis*, Sage.
Signorino, M. (1996) (ed.) *Vent'anni di politica ambientale in Italia*, Maggioli.
Stone, D. (1988) *Policy Paradox and Political Reason*, Harper Collins.
WWF Italia (1996) *Italia 2000. Iniziative per un paese sostenibile*. WWF.

Chapter Seven

Innovation Concepts and Cleaner Technologies: Experiences from Three Danish Action Plans

By Arne Remmen

1. Introduction

In 1986 the Danish Ministry of the Environment initiated a cleaner technology program concerning the development and introduction of cleaner technology in companies. For this purpose some 20 million Danish kroner were spent each year during the late 1980s. With the implementation of two action plans for 1990-92 and 1993-97 the amount spent rose to around 75 million kroner each year.

Cleaner technology is increasingly seen in Denmark as the primary means to reduce the use of resources and the environmental impact of companies – as a first step prior to end-of-pipe solutions. Also "less polluting technology" has moved into the center of policy focus with the implementation of the Environmental Protection Act from 1992.

Since the action plan from 1990 the definition of cleaner technology has been: "that pollution and waste as consequences of the production, use, and disposal of products are eliminated or limited as much as possible so close to the source as possible. This implies that the product or production process is altered so that the accumulated effect on the environment from the cycle of substances and materials is limited as much as possible" (Environmental Protection Agency, 1992, p. 2).

In this paper, I will briefly discuss the changes in the conceptualization of cleaner technology in the three Danish programs. The concept of cleaner technology reflects both problem understanding: what is regarded as a problem, and strategy: how the activities are to be carried out, e.g. the kind of solutions and actors involved.

A critical assessment of the cleaner technology concept can show whether the problem understanding and the strategy have kept abreast with the changes of the activities that have been accomplished. The assessment is based on an analysis of the three action plans (Miljøministeriet 1986, 1990a, 1992), the evaluations made (Andersen & Jørgensen, 1997) as well as reports from central cleaner technology projects and my own involvement in different projects and evaluations.

The development of the concept of cleaner technology can in broad outline be illustrated by distinguishing between three generations of concepts, which partly overlap the three action plans. The activities started in the first generation are, however, to a large extent continued in the second generation, etc. (See table)

1 generation Demonstration Projects	2 generation Diffusion of Cleaner Technology	3 generation Integration
From 1986 90 mio.kr 1986-89	From 1990 230 mio.kr. 1990-92	From 1993 375 mio.kr 1993-97
• Trade survey • Demonstration projects • Information database (RENTEK)	• Handbook of environmental management (- audit) • Environmental management in selected companies • Trade consulting and dissemination • Cleaner technology projects in municipalities and counties • Green public procurements • Life cycle assessment and Eco-labelling	• Integrated quality and environmental management • Employee participation • Preventive environmental and health & safety work in small companies (kr. 80 millions in 1994-97) • EMAS and trade packages (kr. 120 millions 1995-1999) • Activation of network (e.g. training) • Product oriented environmental policy • Demands to suppliers of raw material and machinery • Environmental action plans (companies and local authorities) • Environmental considerations in all sector policies • Differentiated and dynamic regulation

2. First Generation: Technical Demonstration Projects

Primarily three means were applied in the middle of the 1980s, when the Danish Ministry of the Environment implemented the development program for cleaner technology, namely environmental surveys of trades, development and demonstration projects and a compilation of an information system.

The aim of the industrial sector reviews was to give a survey of the most important environmental problems, including a description of where the problems arise in the production process, and how the environmental problems can be technically reduced. Demonstration projects had as their aim to develop and test technical solutions. The rest of the trade is subsequently informed about the results via reports and demonstration days. Finally the information database was intended to hold the available knowledge about cleaner technology and take part in diffusing the concept to firms, authorities, consultants, etc.

The logic of the innovation concept was overview, development, and diffusion. The pivotal point of the concept was the demonstration projects with a demand of testing new ideas and innovations. In these projects primarily engineering consultants and technical managers of the companies were involved. Already at this stage, there was an understanding in the Danish Environmental Protection Agency (DEPA) that in the beginning the focus had to be on demonstration projects and then later more attention should be given to diffusion activities.

This approach tends to reduce prevention to technical fixes: new cleaner technologies can solve the environmental problems. "The engineering mentality" is prevalent, and it is furthermore based on the assumption that when technological solutions are at hand and economically reasonable they will be used. The focusing on "good examples" in the form of demonstration projects also leads to a strategy, which is focused on ensuring the supply of technology, in other words technology promotion.

This can be seen as an example of the classical "technology push" strategy: technological development and diffusion through dissemination of information. The problem understanding and strategy go hand in hand and mutually confirm each other. However, the diffusion of cleaner technology has been considered to be rather slow.

A survey of the trade's environmental problems, technical solutions and a dissemination model consisting of demonstration days, reports and a "not-completed" database are not enough to ensure the diffusion of cleaner technology. The subsequent development does show, however, that these first activities have been sufficient to ensure a continued interest in cleaner technology in Denmark.

2.1 Evaluation of the first generation concept

At the end of the first development program an evaluation of the program was carried out. The evaluation concluded that the program assessed according to its own assumptions had been a success, as the planned activities had been implemented and completed. (Miljøstyrelsen, 1990b, p. 103)

The evaluation contained a critical assessment of the concept in question, which was criticized for:

- exclusively having focused on the production process instead of the products;
- that working environment and technology policy and other sectors than the industry had been given low priority;
- that cleaner technology had been insufficiently incorporated in environmental regulation and administration;
- that barriers to the implementation of cleaner technology had not been examined.

It can be added that the focus has mainly been on cleaner *technical solutions* developed and introduced by experts and this has had the additional effect that the diffusion of the cleaner technologies was limited. Good environmental housekeeping, environmental behavior, changed work routines, education and organizational initiatives have been absent from the involved parties' understanding of cleaner technology and thus from the activities carried out until 1990. (Christiansen & Kryger, 1989, Remmen, 1995).

3. Second Generation: The Diffusion of Cleaner Technologies

The Danish EPA had gradually initiated new projects regarding the critical points mentioned above in the period 1990-93. The effort was expanded to include products, yet primarily life cycle assessment related to product development through the EDIP-project (Environmental Development of Industrial Products). It was stressed that cleaner technology projects must not result in the deterioration of the working environment. The development of a manual of environmental management was completed. Cleaner technology projects in municipalities and counties were implemented. Public procurement was examined as a means. Cleaner technology was stressed in the objects clause of the environmental protection law.

These changes did not take place all of a sudden due to the evaluation, but rather as a gradual change of course based on widespread experience far into the Danish EPA of the fact that the diffusion of cleaner technology was too slow. In other words, the results reached in the early 1990's did not meet the expectations.

The number of means applied has increased markedly during the period. Therefore it is complicated to estimate the problem understanding and the strategy behind the second generation of cleaner technology projects. By and large the cleaner technology projects are continued in the same track with a more prominent focus on promotion and dissemination. The general idea is that cleaner technology must have "help" in order to be diffused.

Especially the four completed *trade consultant schemes* strengthen this impression. The aim of the trade consultant schemes was that consultants with knowledge of the production technology of the trade in question were to analyze the companies' environmental impacts, and then suggest the implementation of cleaner technical solutions that were profitable for the company. The positive result was closer contact to far more companies and also an initial realization of their dissimilarities. The compilation of trade key figures of the average consumption of water and energy and of discharges was also an effective means to reduce consumption and discharge in some of the trades. At the same time the experiences showed that the trades are very different regarding the exchange of knowledge and experiences on cleaner technologies.

Another new means from this period is *environmental management*. In 1992 the Danish EPA published Environmental Management – A Manual of Practical Environmental Work. The main concept was the environmental management circle, which "is to symbolize that the environmental management program is not a finished result, but a continuous process of improving the environmental performance of the company" (Bødker, 1992, p. 11). The manual may be seen as a potential of adding new elements to the cleaner technology concept, especially with regard to the perception of cleaner technology as a principle and a process instead of a specific technical solution.

Earlier manuals from the US EPA like "Waste Minimization Opportunity Assessment" (1988) and "Facility Pollution Prevention Guide" (1992) has been used as handbooks for implementing *cleaner production* in the companies. Many of the same activities are involved: environmental review, planning and organization, targets and action plans, etc. But the old manuals – the first one was translated into Danish – has a linear project approach, while the environmental management manual has a circular process approach (this were also illustrated in the graphic representation in the manuals).

In addition the new manual was based on a comparatively simple approach to environmental management. The more the attention is directed towards environmental management the more apparent it becomes that organizational conditions are a pivotal point of pollution prevention activities in the companies. Cleaner technology in the form of technical experts' adjustments of existing machines may be fine, but a foundation of pollution prevention internally in the company is even more crucial.

The environmental protection law was a clear signal to companies, local authorities and the public that pollution prevention has first priority in the environmental policy. Already the objects clause of the law stresses the intention of promoting the use of cleaner technology. From the comments on the law it appears that the principle about cleaner technology must be "ruling for a holistic and preventive environmental policy" (p. 20). The environmental law gave a more clear authority to the municipalities and the counties about the means that can be used in order to stimulate the implementation of cleaner technologies in the companies.

Earlier the requirements to the companies were solely based on a recipient estimate (i.e. what the environment "can tolerate"). The environmental protection law of 1992 gives a certain legal authority to order a firm to reduce the pollution, when it "results in considerably more pollution, including production of waste, than would have been the case by the use of least polluting technology or best possible purification." (Chapter 5, §41) – i.e. the demands were based on an estimate of what level of pollution is obtainable through the use of cleaner technology.

The enforcement of this regulation has shown that there is a big step from the establishment of the principle of cleaner technology in the law to the adaptation in the municipalities. Many local authority officials and environmental managers in the companies will have to begin rethinking the understandings and strategies.

Also in the promotion of cleaner technology more means are used, which again stresses the increased focusing on diffusion. Information is seen as an important means of overcoming barriers of knowledge and attitudes: Booklets are published about cleaner technology, both a general one and trade specific ones. A video is produced with the message: *Cleaner technology is not a destination but a journey.* A newsletter about cleaner technology has been published since 1993. This is all fine and very necessary. But the question remains whether these initiatives have resulted in a change of the understanding and the strategy behind the cleaner technology projects.

The information efforts of the Danish EPA are extensive with regard to the publication of reports. The extensive production of knowledge within the framework of the action plans for cleaner technology is only rarely followed by an intensive dissemination – this is usually left to the participants of each project. The larger dissemination effort on behalf of the Danish EPA is generally not reflected in relation to the implemented projects. Projects are reported – not disseminated. As a user of the information of the Danish EPA it takes considerable effort to get an overview of projects in progress and reports in the process of being published (Rasmussen and Remmen, 1997).

By the end of 1992 RENTEK, the information database on cleaner technology of the Danish EPA, was launched – after five to six years of development at a cost of a two-figured number of millions Danish kroner. The vision was to create a work of reference containing detailed descriptions of industrial processes and their most important environmental impacts and also regarding the possibilities of cleaner technologies.

The content of the information system is to a high degree a reflection of the first generation concept, whereas new means such as trade-orientated efforts, environmental management and life cycle assessment by and large are absent from the information system. This characterization of the content is not really changed by subsequent updating; just as no decisive new demands are made on the suppliers of information (Rasmussen and Remmen, 1997).

The characteristics of the information effort of the Danish EPA are also reflected in RENTEK. The information database is not an integrated part of the other information activities. Danish EPA supports cleaner technology projects without at the same time drawing the attention to this in the information system. RENTEK is very much an independent project in parallel with other projects.

3.1 Evaluation of the second generation

The second generation of cleaner technology projects continues the activities begun in the first generation, but with a sharper focus on the diffusion of the results from the demonstration projects through experiments with trade consultants and also through an intensified dissemination.

The finished evaluation of the second generation projects has shown that the activities of diffusion and promotion have led to the desired results. "Eight out of ten cleaner technology projects are successful" it appears from The Newsletter on Cleaner Technology. Until the conclusion of the second action plan 80% of cleaner technology projects have ended in entirely or partially useful results, and are put to practical use (Nyhedsbrevet, no. 1, 1995; Andersen & Jørgensen, 1997).

Furthermore, new sectors and trades are involved. New means are introduced, where especially green procurement points towards an initial realization of the significance of the demand-side. Yet it is not a question of a pronounced focus on demand as such, for instance no new general subsidy programs for cleaner technologies are made, even though this has been the case with subsidies to energy savings. The economic resources are still primarily spent on supply – on demonstration projects and on dissemination efforts.

The problem-understanding (the firms must reduce the environmental impacts from the production) and strategy (demonstration projects and diffusion activities) are by and large the same, but with stronger environmental authority as a consequence of changes in the environmental protection law.

However, there is a potential inherent in environmental management that to some extent transcends the cleaner technology concept. Or rather, settles with the fact that cleaner technology has often been regarded as projects rather than a principle and a process. By reducing cleaner technology to projects, then the projects are finished when the solution has been introduced in the demonstration company in question. In many industries a consultant and a company have received support for developing a specific solution to a specific problem without taking much further action. *The project mentality is destructive to the preventive principle stressing continuous environmental improvements.*

The risk related to projects has been reduced by certified environmental management systems where the various manuals and standards stress that the companies are obliged to improve the environmental impacts continuously. Yet the problem of the standards like ISO 14001 is that they stress a top-down management approach, in which active employee participation and the working environment have been given low priority. (Lorentzen, et.all., 1997)

Even though the concept of cleaner technology has not been changed decisively, new activities have certainly been implemented. This implies a more broad understanding of how pollution prevention can be anchored in the companies as well as of the role of local authorities in cleaner technology activities.

4. Third generation: Integration

The challenge to the actors involved is to see cleaner technology as a principle, a method and an attitude, which must be integrated in all parts of the companies, in the inter-organizational network as well as in the environmental regulation and in other policies. There are tendencies of integration in many of the implemented initiatives during the past 4 to 5 years. Therefore a third generation of cleaner technology initiatives can be outlined, which has gone beyond the traditional approach to the projects.

4.1 Environmental management in the companies

During the last couple of years several initiatives have been implemented which are far better integrated into the companies. An integration with the management systems of the company is taking place in the form of support to four projects, which in various combinations have established certified quality and environmental management systems. Two projects have focused on how the employees can take part in the implementation of cleaner technology and environmental management. Environmental management in small and medium-sized companies has been the theme of four different projects.

The step away from projects is most evident in the implementation of the so-called "small" program in 1994 with an annual grant of 20 millions Danish kroner for four years. From this program small and medium-sized companies can get support to hire an employee to take care of the preventive environmental activities including health and safety issues.

Furthermore, in 1995 a support program was established in order to further the introduction of environmental management, including EMAS. The program description stresses a number of principles of environmental management: integration of management systems, active employee participation, the inclusion of the working environment, the connection to product labeling, etc. (Miljøstyrelsen, 1995).

These activities are necessary in order to anchor the preventive environmental activities in the companies, just as environmental management opens up new perspectives in the form of an obligation to continuous environmental improvements – in contrast to the rather static approach in the demonstration projects.

In relation to *the products* the life cycle principle is becoming more widespread: the assessment of the environmental effects of the product from cradle to grave, from raw material to recycling. In Denmark the so-called EDIP project practically had a monopoly in this area. (Wenzel, et.al. 1997). The tools of conducting these assessments do, however, cause methodological problems. At the same time what is needed is a wide selection of different tools, depending on the object of the life cycle assessment.

The requirements of a life cycle assessment vary depending on whether the assessment is to be the basis for setting up criteria for environmental labeling, or whether the assessment is to give a designer producer a quick survey of the environmental impacts of the products. Until now only the quantitative, expert-based approach has received attention. These time-consuming tools can be seen as a main barrier to the diffusion of the life cycle perspective in especially small and medium-sized companies. Other types of tools, e.g. more qualitative and dialogue-based, could contribute to intensify the environmental cooperation between producers and suppliers, and consequently cause environmental improvements in the whole product chain. (Broberg, 1993, Remmen, 1995).

In the winter of 1996 the Danish EPA announced that the future preventive environmental activities must be centered on a product-oriented environmental policy (Miljøstyrelsen, 1996).

The tendency in the activities directed at companies of the third generation of cleaner technology can be characterized as *integration*. The cleaner technology principle is increasingly becoming the foundation of the environmental activities in the companies, even though far from all small and medium-sized companies have found out what cleaner technology is really about in their trade, and even though there are only rudiments of a product-oriented environmental effort in the companies.

4.2 Cleaner technology in environmental regulation

Integration is also an adequate characterization of the intentions of environmental regulation after the coming into force of the environmental protection law. First of all the Danish EPA introduces the fact that the regulation should be *adapted to the situation* depending on the companies' environmental performance and strategies, so that the means applied will be *differentiated* from the environmentally positive over the ordinary law-abiding to the environmentally destructive ones (Goldschmidt, 1993). Secondly, in 1993 the new guide to environmental approval was published, which gives the authorities the possibility to give environmental permits based on what can be reached through the use of cleaner technology (best available technology). The aim is to support this in cooperation with the trade organizations by drawing up and distributing trade information material about the possibilities of using cleaner technology in the production processes of the trade.

However, it takes time to realize the intentions in practice. It takes time to educate the companies to describe the use of cleaner technology in the application for an environmental approval. It takes time to educate the authorities to consider cleaner technology in the environmental approval. And it takes even more time to make the authorities offensive and use their "can" possibilities in the form of making demands on the application and setting up the terms corresponding to what is "least polluting technology" in the trade in question.

A general impression of the *normative* regulation is that it is increasingly characterized by cooperation and dialogue between the companies and the authorities. This is especially reflected in companies that implement environmental management in one form or another.

In addition an increased use of *informative* means is taking place, especially in order to stimulate the companies to implement environmental management, perhaps certified according to ISO 14001 or EMAS. In the same way the Danish law from 1996 about green accounts has an informative aim, where the green accounts can provide a survey of the trade's environmental key figures and in relation to the public increase the interest for the environmental impacts of the companies.

These means can give synergy effects where environmental management, green accounts, etc. can influence the competition parameters on the market as well as contributing to an increased interest and attention from the public (Remmen & Nielsen, 1994).

The environmental debate in the media has to a large extent been dealing with the *economic* means. Green taxes have been made a target of resistance by the large energy consuming companies. But the very debate about this may stimulate some companies to introduce pollution prevention. The green taxes will – all other things being equal – increase the incentive of the companies to environmental improvements.

There are clear tendencies towards an integration of cleaner technology in the environmental regulation with regard to means such as economic (taxes, support schemes), informative (environmental management and green accounts) and normative (supervision, approvals). It may once again be stated that "things take time".

4.3 Cleaner technology and the network of the companies

In the universe of the Danish EPA and consequently in the action plans there have primarily been two to three main actors: the companies, the authorities, and the consultants.

Most of the mediators in the company – authority relation have been neglected up until the mid 1990's. The Industrial Health Service (Bedriftssundhedstjenesten) has the health and safety competence. Technical schools and vocational training centers educate the employees of the industry, who ought to learn about pollution prevention. The energy supply companies are by now able to give advice about the possibilities of saving energy. TIC-centers (Technological Information Centers) are used to advise the small and medium-sized companies. The National Association of Local Authorities in Denmark (KL) and the Association of County Councils in Denmark could take the responsibility of the cleaner technology competence of authorities in charge of supervision and enforcement. Trade organizations and trade unions have contacts to each side of the table and can motivate the companies towards an active effort, etc.

The cleaner technology concept has gradually been integrated in the inter-organizational network. TIC has been responsible for one of the projects about environmental management in small and medium-sized companies. The trade organizations have played a decisive role in the implementation of so-called trade-packages in order to promote environmental management (EMAS) in selected trades, for which 120 million Danish kroner have been appropriated. KL has initiated a number of development projects in municipalities, but training and education of the environmental officials in order to increase their cleaner technology competence has just been implemented even though the Environmental Protection Law was enacted more than five years ago.

Network-based activities can contribute to a synergy effect among the organizations mentioned above such as cooperation on cleaner technology and involvement of new actors in the activities. These institutions each have their contribution to offer in relation to the creation and dissemination of competence and qualifications concerning cleaner technology.

Especially because the barriers of knowledge and attitude are of great importance it seems strange that no one has put any efforts into establishing in-service training for the involved key personnel in the companies, the municipalities, etc. The most odd thing certainly is that the suppliers have had a marginal role. If the suppliers of machines, raw materials and auxiliary materials are involved actively, then there is an evident possibility of taking a short cut to cleaner production processes in the companies.

4.4 Evaluation of the 3rd generation concept

It is conspicuous that two new programs have begun after 1993 to advance the implementation of environmental management. The programs are aimed at small companies and trade-based environmental efforts. These programs are "working programs", in which there no longer is a demand about innovation as in the cleaner technology action plans. In other words, increased *integration* has taken place in relation to the companies and the trades. The tendency towards this integration can also be found within other fields, cf. below:

A general overview shows that in relation to the companies the cleaner technology activities have achieved:

- environmental management and continuous improvements as the pivotal point,
- a wide (uncoordinated) range of tools to support this,

- that employee participation, organizational change and learning processes are more in focus,
- integration in relation to the management systems,
- "working programs" for environmental management (trade-oriented in connection with EMAS as well as towards small and medium-sized companies),
- increasing understanding of the necessity of a life cycle perspective and of cleaner products.

The cleaner technology projects have focused on the production processes and in this regard many companies and their trade organizations now have a platform that enables self-regulation – with the municipalities and interest groups in an initiating and "pushing" role. Still, a more coordinated and disseminated presentation of the tools for cleaner production is desirable, so that the involved parties more easily can find the tools suitable for their conditions and level of ambitions. The next challenge is to diffuse further the life cycle principle and a product-oriented environmental effort in the companies.

Danish EPA and the action plans for cleaner technology have very much played an initiating role for the development outlined above. At the same time the environmental authorities have implemented a number of new means:

- permits and demands must be based on "least polluting technology",
- pressure on the companies about increased "self-regulation" in the form of EMAS and green accounts,
- new normative/informative means such as reference lists, trade information etc.,
- differentiated and adapted regulation depending on the strategies and the performance of the companies,
- green taxes on emissions and resource consumption,
- increased informative efforts about the possibilities of cleaner technology,
- sharper focus on the demand aspect in the form of public green procurement, Eco-labelling, etc.,
- rudiments of a product-oriented environmental policy.

The means of promoting cleaner technology in the production have by and large been introduced. However, neither the Danish EPA nor the National Association of Local Authorities in Denmark have coped with the fact that the local environmental authorities are very different with regard to manning, competence, and interest in cleaner technology. Furthermore, the quality of the dissemination activities leaves much to be desired. In other words, the

Danish EPA has a share of the responsibility for ensuring that the authorities charged with supervision of the companies are prepared to deal with cleaner technology.

The development outlined above has in addition resulted in an increased activation of the network and of interested parties (trade organizations, trade unions, the National Association of Local Authorities in Denmark, TIC, etc.). These organizations and institutions must be far more active and a driving force in order to implement a product-oriented environmental policy. Yet it remains a question how the future division and distribution of tasks and responsibility will be among the parties involved. As pointed out there are "blind spots" such as in-service training efforts, coordinated promotion of a tool package, etc.

The obvious involvement of suppliers of raw materials, machines, etc. in the cleaner technology activities has, on the other hand, to wait for the product-oriented environmental policy.

5. Conclusions: Changes in the Cleaner Technology Concept

The cleaner technology concept has had a relatively dynamic development from predominantly focusing on technical demonstration projects, via activities that had a broader basis directed towards the diffusion of cleaner technology, to integration of the preventive environmental effort in relation to the companies, the network and the environmental authorities. In Denmark cleaner technology has become the "guiding principle" in the environmental policy and is synonymous with pollution prevention in a broad sense.

The change of the concept and the implemented activities are an expression of a changed understanding of the environmental problems as well as of the strategies regarding cleaner technology. Until recently the environmental focus has been on resource consumption, discharges, and waste from the production process, whereas now it includes increasingly the entire production system including the use of products. The change in strategy is from a predominantly technical optimization of the present production process towards a broader effort that stresses environmental management, trade-orientated activities, the involvement of more actors, etc.

Cleaner technology has today become a wide concept covering:

- processes and products,
- employee participation and management commitment,
- environmental management and life cycle assessment,
- green taxes and green accounts,
- environmental awareness and dissemination

- Eco-labelling and green public procurement,
- trade orientation about least polluting technology,
- the authorities as a service-orientated opponent, etc.

From being an effort only involving production engineers and a few external consultants exclusively, most of the relevant actors and stakeholders are today involved in different kind of pollution prevention activities. However, consumers, suppliers and to some extent environmental groups are waiting for "The Cleaner Product Program" in order to be included.

The changes in the problem understanding and in the strategies can be seen as a reflexive learning process where the experiences with one set of means and instruments gradually leads to changes in the understanding of the environmental problems and consequently also in the strategies for implementation of cleaner technology. The most important changes have been:

- from cleaner technology projects *to* dynamic processes of continuous environmental improvements,
- from only engineers involved *to* (nearly) all stakeholders,
- from technical solutions *to* a broad fan of means,
- from cleaner production *over* environmental management *to* cleaner production and products (life cycle management),
- from a "technology push" strategy with the firm as the unit of innovation *to* a mixture of supply and demand strategies with innovation as a distributed process involving many actors and institutions.

References

Andersen, M.S. & Jørgensen, U. (1997) *Evaluation of the Cleaner Technology Program 1987-1992. Results, diffusion and future activities*. Environmental Review, no. 14. Danish EPA, Ministry of Environment and Energy. Denmark.

British Standard: *Environmental Management Systems*. 1992.

Broberg, O. (1993) *Intuitive ingeniører sikrer ikke arbejdsmiljøet*. Loke, nr. 4.

Bødker, J. m.fl (1992) *Miljøstyring – en håndbog i praktisk miljøarbejde*. Miljøstyrelsen, 1992.

Christiansen, K. & Kryger, J. (1989) *Promotion and Implementation of Cleaner Technologies in Danish Industries*. The Environmental Professional. Vol. 11, pp. 199-208.

Elert, C. & Remmen, A. (1994) *Miljøstyringens ABC*. Nærings- og Nydelsesmiddelarbejder Forbundet og Aalborg Universitet.

Environmental Protection Agency (1988) *Waste Minimization Opportunity Assessment Manual*. United States.
Environmental Protection Agency (1992) *Facility Pollution Prevention Guide*. United States.
Goldschmidt, L. (1993) *Myndighederne og virksomhederne: Servicepartner eller serviceorienteret modpart*. Loke, nr. 2.
Kristiansen, K. (1993) *Branchekonsulenter og indførelse af renere teknologi*. Loke, nr. 2.
Lorentzen, B. m.fl. (1997) *Medarbejderdeltagelse ved indførelse af renere teknologi – hovedrapport*. Miljøprojekt nr. 354. Miljøstyrelsen.
Nordjyllands Amt, Miljøkontoret (1992) *Miljøstyring og miljørevision*.
Orientering fra Miljøstyrelsen (1990b) *Evaluering af udviklingsprogrammet for renere teknologi*. Nr. 3.
Miljøministeriet (1990a) *Handlingsplan for renere teknologi 1990-92*.
Miljøministeriet (1992) *Handlingsplan for renere teknologi 1993-97*.
Miljøstyrelsen (1995) *Miljøstyring og miljørevision i virksomheder*.
Miljøstyrelsen (1996) *Debatoplæg til en produktorienteret miljøpolitik*.
Ministry of the Environment (1992) *Environmental Protection Act No. 358 of June 6, 1991*.
Moe, M. (1995) *Environmental Administration in Denmark*. Environmental News no.17, Ministry of Environment and Energy.
Nyhedsbrev 1 (1995) *Renere Teknologi*.
Rasmussen, B.D. & Remmen, A. (1997) *Evaluering af informationssystemet om renere teknologi*. Miljø- og energiministeriet.
Remmen, A. (1993) 'Troværdighed – virksomhederne og miljøet'. *Loke*, nr. 2.
Remmen, A. (1995) 'Pollution Prevention, Cleaner Technologies and Industry'. In Rip, A., Misa, T. and Schot, J. (1995) *Managing Technology in Society. The Approach of Constructive Technology Assessment*. Pinter.
Remmen, A. and Nielsen, E.H. (1994) 'New Incentives for Pollution Prevention. Environmental Strategies for Companies and Public Regulation'. International Association for Cleaner Technologies (IACT), Vienna, April 94. In *Incentives and Policies for Clean Technology, Legislative and Educational Frameworks*. IACT & Federal Ministry of Economic Affairs, Austria.
Wenzel, H., Hauschild, M. and Alting, L. (1997) *Environmental Assessment of Products*. Vol. 1 and 2. Chapman and Hall.

Chapter Eight

Public Engagement and UK Agricultural Biotechnology Policy

By P. van Zwanenberg

1. Introduction

Modern agricultural biotechnologies are viewed by their proponents as potentially revolutionising the food and agriculture industries. Faster rates of growth in both productivity and aggregate productive capacity are envisaged as well as entirely new forms of production. Proponents claim that agricultural biotechnologies will allow more environmentally sustainable forms of production which, for example, could replace petroleum-based inputs to industrial processes or reduce the need to use chemical inputs in intensive agriculture. Broader social benefits are also envisaged, such as more secure food supplies, improvements in food quality and the potential for new and useful products.

Yet many environmentalists and other critics view the entire biotechnology revolution as posing nightmarish risks of the consequences of meddling with nature, inducing further intensification and industrialisation of agriculture, and risking dislocations to existing social and economic relations and moral sensibilities. Research on public attitudes to biotechnology also indicates significant levels of unease and concern among lay publics about unfolding trajectories in biotechnology. (Grove-White *et al* 1997) These also relate, in part, to a range of broad ethical, social and physical risks especially in a context where the motives for innovations in biotechnology are perceived to be solely commercial or for what are seen as trivial benefits (improved flavour, shelf life and so on in the case of food products).

Clearly, sensibilities about the new agricultural biotechnologies can and do conflict. Mindful of both the enormous commercial and political pressure to minimise barriers to innovation in biotechnology and, at least to some extent, of public concerns, government has been anxious to secure the legitimacy of the new biotechnologies. Over the last decade or so a "precautionary" form of regulation has been established and a number of initiatives have been launched to improve public understanding of both the technical aspects of biotechnology and its apparent benefits. Nevertheless, as became even more explicit following the first UK commercial introduction of genetically modified foodstuffs in 1996, public concerns about emerging trajectories in agricultural biotechnology are widespread; indeed research suggests that they intensify the more people learn about developments in the technology. (Grove-White *et al* 1997)

In this chapter we argue that many of the tensions and public concerns surrounding the technology are rooted in the inadequacy of existing political and regulatory arrangements for managing agricultural biotechnologies. Indeed, more generally we would suggest that controversies over the new agricultural biotechnologies are emblematic of a series of problems with traditional expert-based forms of science and technology policy-making in the UK. The chapter explores some of those tensions by focusing on the nature of UK biotechnology policy-making processes and in particular on the ways in which they engage with the wider public through various forms of representation and participation. We argue that, for a number of different reasons, policy has, for the most part, foregone any role in directing agricultural biotechnology trajectories, except in so far as supporting the research agenda of the commercial sector. Indeed, there is a widespread assumption within the policy domain that the management problematique, as far as biotechnology is concerned, is only of one potential physical risks to the environment from individual innovations; risks that are treated as if they were entirely predictable and manageable by suitably qualified experts. Given this official framing of the scope of policy, public participation and representation have been viewed as largely unnecessary. Indeed, the role offered to the public tends to be confined largely to one of consumers; exercising agency, if at all, through choices in the marketplace. This official framing of the scope of policy, and the extent to which it is in conflict with public definitions of what is at stake environmentally and socially in the biotechnology arena, animates a range of different, often poorly understood, tensions.

Since the introduction of commercial products containing genetically modified organisms in 1996, a far wider public debate has nevertheless ensued about agricultural biotechnologies involving actors and organisations in the civic domain, academics, and some food manufacturers and retailers. What is in effect a public technology assessment is taking place. Following the unraveling of the BSE/CJD disaster in early 1996 and the election of a new Labour administration, there does appear to be greater government sensitivity to public concerns about agricultural biotechnologies. Thus far, however, this debate about the wider politics of agricultural biotechnology has made little impact on existing regulatory processes and procedures. It remains to be seen whether and how more innovative cross-fertilisation might occur.

The chapter begins by discussing the opportunities for public engagement and representation in policy for the promotion of biotechnologies.

2. R&D Policy and Public Engagement

UK Biotechnology R&D policy has been concerned primarily with supporting the general biological science base, but in recent years has become far more focused on supporting research trajectories that, under the foresight programme exercise, appear to have potential in leading to commercially viable applications. Indeed, following the failure to patent the British discovery of monoclonal antibodies in the mid-1970s, the internationally renowned strength of the UK science base in molecular biology, the fact that the UK has the largest number of small biotechnology firms in Europe, and the widely held belief that biotechnology represents the next major technological basis upon which economic development will depend, UK biotechnology policy has become dominated by attempts to commercialise publicly funded research and to establish and consolidate relationships between the academic and economic domains. Policy innovations in the last 15 years or so include the formation of 'clubs' in which several firms, together with government, fund promising areas of research and the creation of part publicly owned companies such as Celltech and the Agricultural Genetics Corporation which were set up in the 1980s, principally to transfer biotechnologies from the public sector research base to industry. In 1981 the Science and Engineering Research Council (SERC) set up a Biotechnology Directorate to fund strategic research in biotechnology and promote technology transfer (strategic research applies to an area considered pre-competitive falling in between basic and applied research) In 1994 the research councils were reorganised under an initiative designed predominantly to make public sector research more industrially relevant, and biotechnology now falls under the aegis of the Biotechnology and Biological Science Research Council (BBSRC) whose research trajectories are informed by technology foresight programmes. Whilst the research councils fund basic and strategic research, since 1982, the Department of Trade and Industry's Biotechnology Unit has helped to fund applied scientific research in British industry.

Environmental R&D policy for biotechnology is difficult to distinguish from biotechnology R&D policy in general. Nevertheless, based on the UK government's definition of policy commitments to biotechnology and sustainable development, environmental R&D policy in this arena is confined almost entirely to the development of risk assessment methodology for the release of genetically modified organisms (HMSO 1996b) One exception in recent years was a commitment to investigate policy mechanisms to promote green biotechnologies. But this was subsequently dropped leaving only a programme to publish reports on the application of biotechnology for environmental benefits. (HMSO 1996b, p.76) Under the research councils' clean technology programmes there are, in addition, a number of grants to universities and joint university/industry collaborations concerned, for example with research and development in bioremediation and sustainable crop production systems. (ESRC 1996)

Decisions about the funding of public biotechnology R&D rarely involve direct representation of public interest groups. The SERC Biotechnology Directorate, for example, was run by a management committee, half of whom were industrialists and the other half of whom were academic research scientists. (Sharp 1989) BBSRC research committees do contain one or two consumer and farming representatives but are overwhelmingly dominated by industrialists and senior academics. Public representation is instead viewed largely as the prerogative of Parliament who sanction the general remit of science policy and provide oversight through, for example, various select committees. The 1993 science white paper, *Realising Our Potential*, states that Government funding of science should be driven by two primary objectives, national prosperity and the quality of life. (HMSO 1993, p. 2) Mission statements for government departments and the various research councils reflect those dual objectives. Even though the white paper nowhere defines what is meant by quality of life, the term clearly implies that wider public interests, as well as wealth creation, should be recognised in science policy. Yet, the 1995 foresight panels, comprising of senior industrialists and scientists, and charged with the task of forecasting funding and support trajectories on the basis of the white paper, have to some extent interpreted the quality of life goal as if it were synonymous with wealth creation, at least in so far as there is no recognition that there might be any potential conflict between those two objectives. For example, the Food and Drink foresight panel recommended that modern biotechnologies should be exploited to produce products of use to both end consumers (the public) and intermediary consumers (the food processing industries). Yet although the Panel recognised that there may be a hostile public response to the use of genetic modification technologies in food and drink production it suggested that "[l]imited public acceptance ... is caused by lack of informed understanding" (Cabinet Office

1995, p. 5) rather then any entertaining of the possibility that civic political values might clash with the sensibilities implicit in public and corporate R&D trajectories.

Parliamentary institutions which play some role in overseeing and directing R&D also appear to assume that wealth creation potential is identical to wider public interests. For example, when the Lords Select Committee on Science and Technology reported in 1988 on government research policy in agriculture and food, the committee emphasised the need to ensure that research was responsive to the needs of the agriculture and food industries. Yet, it failed to recognise that there might be a need for any sort of broad social and environmental evaluation built into the determination of public sector research priorities. (Munton, Marsden & Whatmore 1990) Moreover, the same committee, in the course of a 1993 inquiry into biotechnology regulation, received evidence of widespread public mistrust of both government controls and the biotechnology industry. Again this information was interpreted simply as a problem of misunderstanding and even irrationality on the part of the public. (House of Lords 1993, p. 61)

Thus, Parliament, which, in the case of R&D policy, provides the dominant form of public representation has, in practice, failed to articulate civic sensibilities in biotechnology policy. Although there are several reasons underlying this failure, it is worth noting that the commercialisation of the public sector research base over the last two decades has contributed to a diminished potential ability for Parliament to direct and oversee policy. For example, State support for applied or 'near market' research has been gradually withdrawn, in the belief that commercially viable research should be picked up by the private sector, and those science and technology agencies that are regarded as operating near the market have been privatised, or are being considered for privatisation (for example, part of the Cambridge Plant Breeding Institute was recently sold to Unilever). Furthermore, where research continues to be funded within public R&D institutions, government has tried to make research institutions more responsive to the needs of industry by, for example, encouraging technology transfer to the private sector and by constructing internal markets. (Rappert 1995) These factors, together with declining funding for basic science, have placed strong pressures on public sector research institutions to try and hang on to their research teams by making their work relevant to the needs of industry and by trying to attract private sponsorship. (Munton, Marsden & Whatmore 1990) Universities, for example, have been expanding their links with industry, commercialising their in-house expertise through the licensing of technologies to the private sector or developing their own companies and extending market-based principles into research planning. (Webster 1994) All these trends not only underpin a process in which the ideas and problems of industry are transferred to the public science base, but they also hinder public oversight and accountability and they reinforce more general ideological trends that view the marketplace as the proper forum in which to determine technological trajectories.

One of the few attempts, to date, to involve the public actively in discussion of modern biotechnology was the 1994 National Consensus Conference on Agricultural Biotechnology. The conference was organised jointly by the Biotechnology and Biological Sciences Research Council and the Science Museum, largely as a response to evidence of the extensive credibility problems of the biotechnology industry and its regulators in the UK. Notably, however, the conference had no statutory powers to alter policy (as is the case in the Netherlands and Denmark), nor was there any requirement that the legislative or regulatory systems engage with the conference recommendations. Indeed, the chair of the Lords Select Committee on Biotechnology stated at the conference that Parliament would not be influenced by the report. (Purdue 1996)

To summarise, there are virtually no mechanisms or opportunities for direct public representation in policy-making for the promotion of biotechnology R&D. Indirect representation is provided by Parliament but that role is shrinking as a result of the spread of market cultures throughout the science base and where it does exist, the sensibilities of the public are simply patronised as irrational or misinformed. There is effectively no form of representation that in practice adequately conveys civic political values into this part of the policy process. One effect of this disenfranchisement is to displace civic concerns over the direction of biotechnology trajectories elsewhere.

3. Regulatory Controls and Public Representation

Within most sectors of British technology regulation, control is assumed to be primarily a matter for expert policy-making rather than participatory politics. Suitably qualified experts (drawn from government, academia and industry) are brought into advisory committees of various types and these are considered sufficient to represent the wider public interest since the factors that require control are usually portrayed as largely technical in nature and are assumed to capture the sole legitimate ground upon which public concerns may rest. What are seen, officially, as the social dimensions to regulatory decision-making, such as the need to make decisions about the acceptability of risk, are also deemed suitable for expert control, mainly, it appears, because of a deference to the ability of experts to make reasonable hybrid scientific – social decisions. Moreover, the experts themselves and the process of resolving the relevant technical and social concerns are portrayed as independent of sectional interests. Such scientistic pretensions have historically been assumed to confer legitimacy to UK regulatory policy and thus obviate the need for public representation in, or even information about, regulatory deliberations. Yet, in the case of biotechnology, both government and industry have been mindful of, although to some extent also in a state of denial about, deep public anxieties about the credibility of both the biotechnology industry and its regulators and about the wide-ranging social and ethical issues associated with the new technologies. Consequently, the regulatory regime has evolved as a more cautious, transparent and representative beast than is normal for the UK.

For example, the expert committee involved in regulating releases of genetically modified organisms (GMOs) into the environment, the Advisory Committee on Releases to the Environment (ACRE), includes members from trade unions and an environmental group. Information about ACRE's deliberations is made public and opportunities for public comment on proposed releases are encouraged. Details of any applications to ACRE must also be available in the public domain. Furthermore, regulatory controls over the release of GMOs have been designed from the outset to be precautionary (as in the rest of Europe) in so far as they are intended to prevent harm in the absence of evidence of damage. A case-by-case review of each genetically modified organism with a stepped release programme is intended to generate evidence of safety.

Government has not usually seen any need to allow non-industrial sectional representation on any expert committees (the exception here being advisory committees operating under the aegis of the Health and Safety Executive which include trade union appointed experts). ACRE, and the regulations covering the release of GMOs, were established through legislation in 1990. At that time the Green Alliance – a moderate environmental group that plays a crucial role at the interface between government and other environmental NGOs – had lobbied Labour Members of Parliament to amend the proposed legislation in order to include broader representation on ACRE and to allow greater information disclosure. (Levidow & Tait 1993). The Government conceded, expanding the composition of ACRE and providing for a public register to be set up detailing all release notifications, subsequent advice and decisions, and requiring firms applying for a release to advertise that fact in the local press. Broader representation was supported by some industrial and trade union representatives on the grounds that it would be preferable to contain public unease within the committee rather than in the press. (ibid.) Moreover, some industry representatives have also emphasised the need to win public confidence and avoid, for example, the risk of an appeal against ACRE's advice.

Regulatory controls over agricultural biotechnology are split, however, between environmental, agriculture and food issues. Whilst controls over the release of GMOs have evolved in ways that are more representative and open than is usual agricultural products derived from genetic modification processes are regulated in a more traditional way by the Veterinary Products Committee (VPC) which consists entirely of experts drawn from government and academia. Deliberations of the VPC and their advice to Ministers remain entirely confidential. The Advisory Committee on Novel Foods and Processes (ACNFP) regulates food products derived from genetic modification processes and falls in between the VPC and ACRE in terms of openness.

Framing Risk

ACRE uses a technical risk assessment to evaluate each proposed release and then advise the Department of the Environment on whether or not to grant a consent and if necessary what conditions must attach to that consent. ACRE, as indeed is the case for all the other relevant advisory committees, is restricted to addressing the potential physical impacts of the single product in question assuming in effect that these are the only significant parameters of risk associated with agricultural biotechnology as a technological system. Wider considerations such as the overall social and political implications of reliance on GMO technology are excluded from consideration as indeed are cumulative ecological implications, such as for biodiversity. Similarly, within the boundaries that restrict deliberations to each single GMO, dimensions of risk such as the justification and need for the product, the benefits it may involve, the context in which it will be used (such as the extent to which official regulations can be adequately enforced in practice) and the capacity to ensure decisions are reliable in practice (for example measures for post-market monitoring), are all officially excluded. Nevertheless, it is likely that those wider issues, in practice, do get discussed at the margins within expert committee deliberations even though there is no overt recognition that they should inform regulatory decision-making.

Given that the regulatory system is the only fora for public debate, the wider concerns of the public and public interest groups, which may, and do, relate precisely to those dimensions of risk that fall outside regulatory remits, have no route of expression.

Parliament could have a role in providing a forum for public debate on those wider issues, and indeed ACRE members themselves considered these types of questions as a matter for Parliament rather than regulators. (Mayer *et al* 1996; Levidow & Tait 1993) In practice, however, this role has not been taken by Parliament. As discussed in the previous section, government has not perceived any need for wider technology assessment and has provided no forum for discussion of such issues. Indeed Parliament has largely failed to even interpret public concerns as legitimate, assuming these to instead be rooted in misunderstanding of the technologies' benefits and in public irrationality.

In addition to the exclusion of wider risk issues, a more subtle framing process occurs within the regulatory process in which the pertinent technical risk questions tend to be defined as involving only those that are in practice amenable to controlled scientific assessment. In part, this stems from a general culture of "sound science" within British regulation that tends to reduce complex open-ended and intractable scientific questions to precise scientific parameters amenable to unambiguous assessment. Within the biotechnology arena such reductionism is apparent, for example, in ACRE's definition of the environment as the *natural* environment and not that affected by prevailing agricultural practices. (Wynne and Mayer 1995) Thus in assessing potential risks to the environment, ruled out of consideration, for example, are possible risks arising from the characteristics of a herbicide resistant GMO and contingent land-use practices such as herbicide usage. Such considerations underpinned precisely divergence's between British regulators and those of Norway, Finland, Sweden, Denmark and Austria over an assessment for commercial clearance of a herbicide tolerant oil seed rape in 1994. (Levidow *et al* 1996) The latter countries were concerned, for example, about the possibility that gene transfer from the oil seed rape to relatives might pose risks of multi-tolerant weeds developing and consequently becoming an agricultural problem requiring greater use of herbicides. In the British case, transfer of the herbicide resistant gene was not considered a problem largely because ACRE's definition of the environment did not include agriculture. Instead, responsibility for herbicide-use lies with the Ministry of Agriculture, Fisheries and Food (MAFF) but MAFF have argued that uncertainties in predicting the ways in which pesticide usage varies already would make it too complicated to predict how herbicide-resistant GMOs would affect herbicide use. (ENDS 1994) Thus, risks identified by other European regulators were defined out of consideration by UK framing commitments.

Similarly, one potential problem of developing and using herbicide resistant crops is that the herbicide resistant crop may reappear as a volunteer, i.e.

effectively a herbicide-resistant weed in different crops. Yet due to ACRE's framing of risk this is not necessarily considered a problem. ACRE assume that additional herbicides and management practices could be used to control any such problem. In other words the assumption is that risk is defined, in this context, as economic harm and not as resulting, say, solely from the need to use addditional herbicides. The point here is that what actually gets defined as a risk cannot be read deterministically from empirical evidence but is instead contingent on human choice and value.

A further source of indeterminacy, and one that is encountered in almost all attempts to assess the risks of novel products in open-ended real world conditions, is the fact that risk information is generated from laboratory and field studies and must then, at some point, be extrapolated to commercial use scenarios. The very factors that render laboratory and field trials useful – namely that some forms of control can be imposed on experiments in order, for example, to identify causal relationships or measure key parameters – do not usually exist in real-world conditions. When extrapolating to open-ended commercial use scenarios it has to be assumed that the conditions critical for ensuring minimal risk will be replicated in less controlled conditions. This is a major source of uncertainty. The resulting knowledge is conditional on the assumption that all the key (possibly multiple) causal relationships and contexts are known and can be and have been examined under controlled conditions. Yet, the conditionality of knowledge generated in this way is rarely acknowledged by regulators.

What the above examples and points illustrate is that the current approach to dealing with biotechnology regulation effectively elevates the symbolic role of risk assessment in so far as risk issues are treated as if they could always be objectively and reliably assessed. Yet, by insisting that risk assessment is not subject to at least some degree of social choice, this disenfranchises wider public engagement and debate.

To recapitulate briefly, as research on public concerns about biotechnology indicate, there exist, in the case of biotechnology regulation, competing civic and official/corporate definitions of what is at stake environmentally in the biotechnology arena. Furthermore, the framing of risk within the regulatory sphere results in the inability of the public – policy interface to allow civic framings of the relevant risk issues to be debated or engaged with in any way whatsoever. The regulatory system also tends to reduce complex risk problems in such a way that many uncertainties are defined out of existence, indeterminacies are unrecognised, and which then insists that this form of reductionism is not open for negotiation or challenge. All this underpins a series of deep tensions within this particular technological arena, as well as posing several dilemmas for public engagement

Tensions & Dilemmas

Firstly, many environmental groups, until recently, have had minimal activity in the regulatory GMO domain, largely because they regard their own wider risk concerns as having no route of expression. Even within narrow technical questions of risk, environmental groups have rarely submitted comments to ACRE, in part because they believe that their safety concerns would be disregarded as a consequence of the exclusion of key uncertainties by the committee's framing commitments. (Levidow 1996) Indeed participation in the risk assessment process presents a dilemma for environmental groups, as is recognised by key actors in the relevant public interest groups, (ibid.) in so far as if their arguments are adapted in such a way that they are consistent with regulators' reductionist framings of risk, this may marginalise their broader concerns and also lend legitimacy to official framing commitments. Environmental and consumer groups nevertheless feel that genetic engineering is a critical issue but their minimal overt engagement with the policy process may create the mistaken impression in official and corporate domains that there is public approval and acceptance of agricultural biotechnology controls and developing technological trajectories. (Mayer *et al* 1996)

At the same time, however, a second tension exists in so far as there may be pressures on the regulatory system to act on wider public concerns about the technology that are excluded by the regime's remit. In part, this expectation has been encouraged by government in the ways in which the procedures and composition of ACRE have been established but also because government has blurred the fact the ACRE's remit in practice excludes the wider implications of GMO use. For example, in reply to Labour Members of Parliament's proposals to establish an ethics commission on GMO releases, government exaggerated ACRE's remit, suggesting that such matters should be more appropriately taken by ACRE. (Levidow & Tait 1993, p. 208)

A recent controversy illustrates the tension between public expectations of participation in ACRE's deliberations and actual practice. The case concerned one of a series of experimental releases of a viral insecticide – known as a baculovirus – modified to contain a scorpion venom gene and intended to control caterpillars in cabbage. ACRE reviewed the experiment and cleared it in 1994. Objections were raised and lodged with the Department of the Environment by several members of the public (in this case, scientists living in the vicinity of the proposed release) who were concerned that the baculovirus might spread to affect caterpillars outside the experimental release site and because of the risks that the scorpion gene could transfer to other viruses. (Wynne & Mayer 1995)

The objectors were told, however, that their concerns were too late because the application was a repeat license and had entered a fast-track procedure. Levidow (1996) suggests that the DoE had no means to satisfy objections to the regulatory process because it assumed that ACRE's broad composition would be sufficient to substitute for such involvement. Under pressure, ACRE later reconsidered the application and again cleared it, arguing that the safety judgment in this particular case related to the experimental design of the release and not the nature of the GMO *per se*. In other words risk decisions were based on the specific confinement conditions that applied to the release (which were thus assumed to prevent any risk to the environment) without having to engage with the more complex issues posed by the characteristics of the GMO and the potential for gene transfer.

Yet paradoxically the experiment was intended to examine whether the baculovirus could invade native populations if used commercially even though the confined conditions of the field trial precluded the conditions required for transfer. (Wynne & Mayer 1996) In this particular case, the objectors were effectively questioning the entire R&D trajectory and asking how the risks of gene transfer could be determined prior to relaxing the confinement conditions. Although a reasonable question, it did not enter the remit of decision-making about the risks of the individual baculovirus experiment even though at some later point ACRE will have to extrapolate data from the field trial to the conditions of full commercial use.

A third tension, also related to the absence of a forum in which the broader social implications of biotechnology can be debated, is that public interest groups and the public more generally may adjust for their lack of effective agency by focusing, if at all, on particular products of genetic engineering or single risk assessment controversies. (Grove-White 1996) Individual products may provide, in other words, a surrogate and tangible object from which to express and attempt to debate the technology as a whole. Such heightened incidences of public anxiety, apparently focused on the properties of a single product, have occurred in other areas of regulation (for example over single pesticides such as Alar in the late 1980s and animal welfare issues in the 1990s). These explosions of public anxiety may well be interpreted by industry and regulators as further evidence of the irrationality and misunderstanding on the part of the public rather than a reflection of the structure of public policy.

A fourth and somewhat complex tension concerns the high degree of (latent) mistrust of, indeed alienation from, existing biotechnology regulatory institutions expressed by lay publics. (Grove-White *et al* 1997) This tension is problematic for several reasons. Firstly, the reasons for such mistrust are generally misconstrued by government and regulators. Sociological research on public responses to science and expertise suggests that public risk perceptions relate not so much to the evaluation of physical risks but are based, just as rationally, on judgments about the behaviour and competence of expert institutions. Lack of trust is thus a reflection of peoples experiences and perceptions of regulators, including, for example, past experience of regulatory behaviour, the extent to which framing commitments obliterate wider public concerns, the realities of how regulatory institutions respond to open-ended, indeterminate scientific problems and the extent to which there are procedures in place for ensuring proper official behaviour. (Wynne 1996) Yet government has responded to survey evidence suggesting that the public are wary of government and industry controls by emphasising the need to educate the public about the 'real' risks and benefits of the technologies. (House of Lords 1993, p. 50) Some of the more thoughtful regulators and officials also acknowledge the need to ensure transparency in regulatory deliberations as a necessary basis for ensuring trust yet even this somewhat more progressive stance takes it for granted that there is public trust associated with the way in which risk is framed by the regulatory system and the ways uncertainties are handled.

Secondly, public mistrust is frequently exhibited in practice as far more ambivalent feelings towards official controls. There is, for example, an obvious lack of overt dissent or opposition to biotechnology in the UK despite evidence that lay publics place far greater trust in environmental groups than they do in industry and government. (see, for example House of Lords 1993, p. 50) Again sociological research suggests that realism on the part of lay publics about both the extent to which people are fully dependent on expert judgments and their effective lack of agency in the process of official assessment of risk results in a "virtual" trust. (Wynne 1996) This poses a problem for regulators as this form of "virtual trust" is not rooted in an appreciation that regulators are adequately taking into account public sensibilities. In the event, for example, of mistakes or identifiable harm arising from the release of GMOs, 'virtual trust' is likely to rapidly disappear. Indeed, the narrow reductionism of GMO regulation and thus official ignorance of key uncertainties in the risk assessment process, means that regulators effectively deny their own responsibility that any mishaps might at some point occur. (Grove-White *et al* 1997) Patronising the public by insisting that safety is an easily obtainable commodity is likely to backfire. There are clear parallels here with the BSE saga in the UK in so far as the relevant regulatory authorities categorically denied the possibility that the cattle disease posed a risk to human health despite gross scientific uncertainties about such a relationship. Public trust in that regulatory regime has subsequently rapidly diminished.

4. The Experience and Strategies of Civic Engagement

For the reasons described above, environmental NGOs have had little formal involvement in the biotechnology policy arena. The scope for NGOs to influence policy seems therefore to be confined to the general political domain and via actors outside of formal UK policy mechanisms. Yet, there has, until recently, been relatively little lobbying and campaigning work outside of formal policy processes. In part, this is due to the difficulties of making agricultural biotechnologies into a political issue, given the absence of official fora around which to hang the issues and, more pertinently, the fact that most technological activity has been confined to the R&D sector. GMO products, until very recently, have not become fully commercialised.

In so far as campaigning and lobbying activities have occurred prior to 1996, these have been conducted primarily by the smaller NGOs. These include, for example, the Food Commission, the Green Alliance (a lobbying group whose Parliamentary Officer became the 'expert environmentalist' represented on ACRE), and the Genetics Forum, a very small group concerned only with biotechnology. The Genetics Forum provides information about experimental releases of GMOs through its newsletter, the Splice of Life, and it provides information to journalists, but the Forum is small – only two part-time workers are employed. Since 1993 Greenpeace has funded the Genetics Forum in order that it can monitor the public register. (Levidow 1996)

Since 1996, however, several GMO products began entering full commercial use at which point several of the larger campaigning environmental organisations began working on GMO issues. For example, in late 1996, modified soya began to be imported into Europe for the first time. Greenpeace initiated a European-wide campaign on GMOs, blockading shipments in several sites in Europe. Greenpeace and other NGOs have also been active highlighting some of the uncertainties in scientific knowledge about the consequences of releasing GMOs, (Greenpeace 1997) and, in conjunction with parts of academia, disseminating the wider risk concerns that, historically, have been ignored by regulatory institutions. (Grove-White et al 1997). Recent controversies throughout Europe concerning, for example, the introduction of Novartis' genetically modified maize and the labeling of GMO products have also helped to focus official and commercial minds on the lacuna in public confidence in the technologies and existing regulatory arrangements – a development that has greatly intensified in the UK in the wake of the BSE/CJD saga.

In the case of Novartis' genetically modified maize, safety concerns raised by the Advisory Committee on Novel Foods and Processes – in relation to a marker gene conferring resistance to the antibiotic ampicillan – were overruled when the European Commission sanctioned the product throughout the Union. At that point Austria and Luxembourg temporarily banned the maize on ecological and human health grounds, actions which the Commission then threatened to overrule. Such controversies can only cast doubt on official claims that GMO related risks always can be and are assessed reliably and objectively. They also raise suspicions that commercial considerations – in this case EU trade relations with the US – may sometimes take precedence over precautionary approaches to risk assessment.

In the case of labeling, Monsanto began importing modified soya without separating it from non-modified soya, claiming that the two products were 'substantially equivalent'. Concerns about labeling were raised not only by environmental and consumer groups but also by some manufacturers and retailers. Indeed, the Food and Drink Federation, an industrial NGO argued in favour of labeling and one major retailer announced that it had managed to source non-modified soya and that none of its manufactured food would contain the modified product. Again, explicit conflicts between different branches of the food industry, coupled the prospect of commercial actors feeling obliged to pre-empt regulatory decision-making, is unlikely to reassure publics sceptical about existing regulatory arrangements.

More recently, there are indications that government is attempting to address some of the concerns that thus far have been ignored by the policy process and may perhaps open up new opportunities for public engagement. For example, in 1996 the UK Government's Panel on Sustainable Development argued that current regulatory frameworks were inadequate to deal with wider public concerns and this led to a National Biotechnology Conference held in the spring of 1997. The conference recommended that new mechanisms be established to deliberate wider dimensions of risk in relation to biotechnology. In addition the Ministry of Agriculture, Fisheries and Food decided, for the first time, to consult widely on its decision whether to license use of a genetically modified oil seed rape by farmers.

It is, as yet, difficult to predict how government will follow up its tentative attempts to engage with and understand wider public sensibilities. The new Labour administration appears to be far more sensitive than its predecessor to public attitudes, although the political and economic pressure for maintaining a maximum degree of freedom to innovate remain extremely powerful. Furthermore, some of the very factors that are contributing to contemporary concerns (globalisation of regulation, the growth of market cultures, national deregulation, and the consequent diminution of the capacity for state direction and oversight) also mean that it may be extremely difficult for national governments to restructure regulatory and political control of the technology.

5. Summary

To summarise, policy for the promotion of agricultural biotechnology has been concerned almost exclusively with promoting the competitiveness of the domestic biotechnology industry. Conceived of in this way, there has been little acknowledged need to allow wider public participation in this aspect of policy. Indirect representation is provided through Parliament but that body has effectively assumed that the public interest is synonymous with wealth creation; an assumption that is not borne out by research on public attitudes to agricultural biotechnology. Policy for the regulation of agricultural biotechnology, on the other hand, is almost entirely concerned with physical risks, risks that are portrayed as fully subject to reliable, objective analysis and are thus assumed to be the sole province of experts. Neither aspect of policy allows deliberation on any wider dimensions of risk, nor has that role been taken by other institutions such as Parliament. As such, the policy regime effectively takes for granted a wider social consensus about the desirability of the technology as a whole, the particular trajectories of biotechnology innovation that have emerged, and the existing technological structure which current biotechnology trajectories reinforce.

Although there are some limited forms of civic engagement in regulatory decision-making processes these have been structured in such a way that it has been difficult for environmental and civic NGOs to even articulate their wider concerns let alone challenge and open up expertise or reframe dominant approaches to managing agricultural biotechnologies. Civic resistance or disquiet to emerging trends has tended to be dismissed by government and industry as irrationality (and policy responses to such tensions have tended to be based on that assumption) rather than by entertaining the possibility that public concerns might, at least in part, be bound up with existing institutional and procedural practices and approaches to directing and regulating the technology in the UK. These tensions have pervaded the arena for the last few years and as some official and corporate actors are belatedly beginning to recognise, arguably require policy and institutional innovation at a time when the products are already entering into commercial use.

References

ACOST (1990) *Developments in Biotechnology*, Advisory Council on Science and Technology, Cabinet Office, HMSO.

Ashford, T. (1996) 'Regulating Agricultural Biotechnology: Reflexive Modernisation and the European Union', *Policy and Politics*, Vol. 24, No. 2, pp. 125-135

Blowers (1987) 'Transition or Transformation? - Environmental Policy Under Thatcher', *Public Administration*, Vol. 65, pp. 277-294.

Brunner, E. (1990) 'Science, Secrecy and BST', in P. Wheal & R. McNally (eds) *The Bio Revolution: Cornucopia or Pandora's Box*, Pluto.

Cabinet Office (1995b) *Foresight No. 7, Food and Drink*, Office of Science and Technology, HMSO.

ENDS (1994) 'Genetically Modified Rape Blazes Trail for Industry and Regulators', *ENDS Report*, No. 239, December 1994.

Elzinga, A. & Jamison, A. (1995) 'Changing Policy Agendas in Science and Technology', in S. Jasanoff *et al* (eds) *Handbook of Science and Technology Studies*, Sage.

ESRC (1996) *Directory of Clean Technology Research*, HMSO.

Greenpeace (1997) *Genetic Engineering: Too Good to Go Wrong?*, Greenpeace.

Grove-White, R. (1991) *The UK's Environmental Movement and UK Political Culture*, Report to EURES, November 1991, Centre for the Study of Environmental Change, Lancaster University.

Grove-White, R. (1996) 'Environmental Knowledge and Public Policy Needs: On Humanising the Research Agenda', in Lash et al (eds) *Risk, Environment and Modernity*, Sage.

Grove-White, R. et al (1997) *Uncertain World: Genetically Modified Organisms, Food and Public Attitudes in Britain*, Centre for the Study of Environmental Change, Lancaster University.

HMSO (1993) *Realising Our Potential: A Strategy for Science, Engineering and Technology*, Cm 2250, HMSO.

House of Lords (1988) *Agriculture and Food Research*, Select Committee on Science and Technology, 13-I, HMSO.

House of Lords (1993) *Regulation of the United Kingdom Biotechnology Industry and Global Competitiveness*, Select Committee on Science and Technology, 80-I, HMSO.

Levidow, L. (1996) *UK Safety Regulation of GMO Releases: The Role and Limits of Public Participation*, workshop paper for meeting on Transparency and Public Discourse in the EU in the Area of Genetic Engineering, 9-10 May, Brussels.

Levidow, L. & Tait, J. (1991) The Greening of Biotechnology: GMOs as Environment-Friendly Products, *Science and Public Policy*, Vol. 18, No. 5, pp. 271-280

Lowe, P & Goyer, J. (1983) *Environmental Groups in Politics*, Allen & Unwin.

Mayer, S. et al (1996) *Uncertainty, Precaution and Decision Making: The Release of Genetically Modified Organisms into the Environment*, ESRC Global Environmental Change Programme Briefings, Number 8.

Munton, R., Marsden, T. & Whatmore, S. (1990) 'Technological Change in a Period of Agricultural Adjustment', in P. Lowe, T. Marsden, & S. Whatmore (eds) *Technological Change and the Rural Environment*, David Fulton.

Rappert, B. (1995) 'Shifting Notions of Accountability in Public and Private-Sector Research in the UK: Some Central Concerns, *Science and Public Policy*, Vol. 22, No. 6, pp. 383-390

Sharp, M. (1985) *The New Biotechnology: European Governments in Search of a Strategy*, Sussex European Paper No. 15, Science Policy Research Unit, University of Sussex.

Sharp, M. (1989) 'Biotechnology in Britain and France: The Evolution of Policy', in M. Sharp & P. Holmes (eds) *Strategies for New Technology: Case Studies from Britain and France*, Philip Allan, Hemel Hempstead.

Shohet, S. (1996) 'Biotechnology in Europe: Contentions in the Risk-Regulation Debate', *Science and Public Policy*, Vol. 23, No. 2, pp. 117-122.

Webster, A. (1994) 'UK Government's White Paper (1993): A Critical Commentary on Measures of Exploitation of Scientific Research', *Technology Analysis and Strategic Management*, Vol. 6, No. 2, pp. 189-201.

Wynne, B. (1982) *Rationality and Ritual: The Windscale Inquiry and Nuclear Decisions in Britain*, BSHS.

Wynne, B. (1996) 'May the Sheep Safely Graze? A Reflexive View of the Expert-Lay Knowledge Divide', in Lash et al (eds) *Risk, Environment and Modernity*, Sage.

Wynne, B. & Mayer, S. (1995) 'Evaluating the Consequences of Releasing Genetically Modified Organisms', in Sandberg, (ed) *Release and Use of Genetically Modified Organisms: Sustainable Development and Legal Control*, Proceedings of a Conference Organised by the Norwegian Biotechnology Advisory Board, 13-14 September 1995, Oslo, Norway.

Chapter Nine

Constructive Technology Assessment Comes of Age
The birth of a new politics of technology[1]

By Johan Schot

> "*Romanticism as an alternative to the machine is dead (...) but the forces and ideas once archaically represented by romanticism are necessary ingredients in the new civilization, and the need today is to translate them into direct social modes of expression, instead of continuing them in the old form of an unconscious or deliberate regression into a past that can be retrieved only in phantasy ?*"

This article will address the content and nature of the direct modes of expression that Mumford sought. To me these are the efforts of Constructive Technology Assessment (CTA). CTA can be viewed as romantic insofar as it brings to life the sociotechnical critiques and practices which were expressed by certain Romantic counter-movements in the early nineteenth century.[3]

The core of the CTA perspective is the idea that the social problems surrounding technology can and must be addressed through the expansion of the design process.[4] This expansion implies the involvement of societal actors, particularly those who experience the effects of evolving technologies but are not actively involved in the development of the technologies. They can be consumers, citizens, employees, unions, environmental movements and cities.[5] Design should not be viewed merely as the first phase of the innovation process, but as an crucial aspect of technical change.

An example of CTA. The development and application of modern biotechnology is controversial. Especially the genetic modification of animals and the related applications in the food industry raise social questions. Questions are posed by environmental and consumer groups and relate to public health, environmental effects and ethics. CTA would involve the organization of a set of activities whereby the social issues are articulated and coupled to the design process. The activities can take the form of experiments for example with genetic screening[6]; or dialogue workshops, consensus conferences (public debates), scenario workshops, or citizen reports. The crucial feature for labeling such activities CTA is to determine whether it has a connection with a design process. If activities are organized to inform the public, for example, without any intended or unintended consequence for design, it cannot be seen as CTA. CTA is explicitly taken up in Denmark and the Netherlands. Here CTA is a departure from (traditional) technology assessment, which is aimed at mapping the effects of the given technological options, and not at influencing and broadening the design process. CTA must not only to seen as a new form of technology assessment only. Proliferation of CTA activities will result in a new politics of technology.[7]

CTA philosophy and practice are on the move. This article is an attempt at stock-taking. Now that the first studies have been completed, it is useful to evaluate the state of CTA.[8] Such an evaluation is the starting point for further development. In this article I will evaluate CTA from a longterm perspective. This will make it possible to highlight the underlying CTA concern to establish a new politics of technology. Such politics will feature three qualities: anticipation, reflexivity and social learning. Its normativity rests on these features.

CTA practices may be viewed (romantically) as a new form of politics which replaces the problematic modernist manner of managing technology. CTA is thus a critique of current "management practices".[9]

I will begin by presenting some snap shots of the modernist management of the Machine, as Mumford had it. In the western world modernist management came to dominate beginning in the late eighteenth century. The examples illustrate a pattern of technology politics in a particular period. The chosen snap shots are episodes of resistance to technology. Resistance is interesting for it reveals what can and what cannot be done and said. In this context I will explain the efforts of CTA and enumerate its features. These features provide the criteria for evaluating current CTA design practices. In conclusion I will raise the issue whether the proliferation of CTA practices will lead to the democratization of technological development.

First Snap Shot: The Luddites

In the early modern period a distinct technological domain did not exist. Technological development was embedded in religious, economic and social practices. Technological development was assessed against social norms. This took place in guilds, for example. They often slowed specific innovations. It should be pointed out that guilds were not against all forms of technological development; they hindered the development of only those technologies which were contrary to their ideas about the "good society". Technological development was influenced by more than the regulatory (and evaluative) practices of guilds. It also was shaped by a variety of protests, some of which were institutionalized. Examples include organizing demonstrations, taking up petitions and threatening inventors and entrepreneurs. Breaking machines also was a form of protest.

The destruction of machines is often associated with the acts of the Luddites, the English workers who destroyed textile machines in the early nineteenth century.[10] Until the 1960s the Luddites were held up as technophobes. Historians viewed them as the victims of progress, who saw no other recourse than taking out their blind aggression on the machine. Often it's added that every new technology is analogously misunderstood and resisted, but that resistance eventually subsides. Meanwhile this image of the Luddites, thanks to the research by E.P. Thompson (1963) and E. Hobsbawn (1952), has been corrected. According to their research, since the 17th century organized machine-breaking was a rather popular and successful form of protest. It was better than striking, because employers couldn't employ scabs to keep the machines in operation. More importantly, machine-breaking was not based on a disdain for technology in general. On the contrary, it was directed at particular machines. The only machines destroyed were the ones against which the workers had particular grievances, as, for example, those that reduced the quality of life. Other machines in the factory, which were not seen to have unacceptable effects, were left unscathed.

The Luddites' resistance ran deeper than that against particular machines. Theirs was resistance to the rise of a new kind of society whereby management had the right to introduce machines which made workers redundant, producing unemployment and lowering the quality of the products and thereby the quality of life. To management the Luddites were criminals. Initially the Luddites had English law on their side, for machine-breaking as a form of protest was legitimate. Luddites were not alone in their dissent. They were supported by craftsmen, small-time entrepreneurs and conservative politicians, the last of whom were strongly influenced by Romantic authors as Carlyle and Southey.

Second Snap Shot: Resistance to Elevators

One of the few outbursts of resistance to the machine in the Netherlands took place in the early twentieth century. In 1905 grain elevators were introduced at the Rotterdam harbor (Van Lente, forthcoming). The elevators were large sucking devices which conveyed grain from one ship to another without human intervention. Thousands of dockworkers, who had been carrying sacks of grain from one ship to the other, were to be thrown out of work. Massive resistance emerged, with some success. For a number of years the elevators remained dormant, and eventually some dockworkers received new tasks and higher wages. Strikingly, the elevators did not come under discussion. Union leaders swore by new technology and said it would be bring only good to the harbor. Damaging machines were not seen as an option; it was condemned as Luddism. Thus Luddism had become a term of abuse.[11] The protest revolved around who would profit from the novelties, the laborer or the capitalist. That kind of reception was wide spread in the Netherlands at the time. The dominant conviction was that "technology cannot and must not be resisted" (Van Lente 1988).

Third Snap Shot: The Schiphol Affair

In 1969 the director of Schiphol (Amsterdam) Airport made a plea for a drastic broadening of the airport (Schot 1995). A long battle ensued between the national government, the provincial government and various local municipalities, all against the backdrop of the organized resistance by a variety of local communities and environmental groups. The issues finally were resolved in February 1995 with a decision by the Kok administration to allow the construction of a fifth runway. Schiphol would be allowed to grow, within certain limits. The number of residences to be effected by serious noise pollution was set at a maximum of ten thousand. This decision was guided by a project group of the directly affected, who arrived at that consensus before the formal decision-making procedures commenced, including a public inquiry. The so-called "dual decision", taken within certain parameters, has become a precedent for all discussions. Since then the discussions have not concerned whether Schiphol should grow, but which type of runway would ensure steady growth with the least noise.

As I've argued elsewhere the drive to expand cannot be understood merely in terms of economic interest. The broadening of Schiphol is a part of the story surrounding the Netherlands as "Distribution Country". The broadening is a way to relive the Golden Years, when the Netherlands and especially Amsterdam were the hub of international trade. Resistance to growth was viewed as resistance to the Netherlands as Distributor and thus as resistance to progress and a sound economy. Another reading of the Distribution Country was forgotten, however. This would be an emphasis on transport with added value. Effort would be made to attract to Schiphol passengers bound for the Netherlands only. Now Schiphol's aim is directed at maximizing transit passengers.

Environmental groups and local communities hammered home the adverse environmental effects and trivialized the national economic interest. They called for stricter norms and limits to growth. At the same time they attempted to develop alternatives. For example, they made a plea for Schiphol to be developed into a "railport", whereby passengers bound for Frankfurt, Paris and other European cities would be transported not by plane but by rail. Schiphol has incorporated this idea into its plans, but without having opened a discussion about the future of the airport. Future plans are dictated by the growth in the number of flights and the competition between European airports.

Two Centuries of Resistance

What do we notice if we compare these three histories of resistance to specific technologies? Luddite resistance was directed against the scenario of industrial civilization implied in the coming of the new textile machines. The resistance did in part have the aim of retaining the existing order. Wittingly or unwittingly, the protesters also came to formulate new frameworks within which technological development could be assessed. The Luddite movement was knocked down, but their wish to assess and construct technologies led on into Owenism with its alternative technology workshops and finally the work of many social movements (see Eyerman and Jamison 1991). The dominant way of addressing societal problems embedded in new technologies became, however, regulation. During the course of the nineteenth and twentieth centuries various forms of regulation came into being relating to technological development. Environmental regulation, public health regulation, etc. One feature of the developments is however, that the technology itself is not called into question. Also no alternative technology path are constructed. Rather, the issues involve ex post compensation and repair. The importance of compensation was apparent in the elevator affair. The introduction of elevators was viewed as unavoidable. The protest related to the retention of wages and employment. Having a say in the development of the technical infrastructure of the harbor was not regarded as a sensible activity for labor. The resistance to Schiphol also concerned setting limits to the promotion of new technology (or in this case a large technical system). This time alternative ideas were developed, but they didn't resonate greatly. There was not an institutional or rather negotiating arena for the further development of the common ideas. Constructive interaction between proponents and opponents was lacking.[12] Willingly or unwillingly all those involved came to play the role of proponent or opponent.

The CTA Perspective

In my view the efforts of CTA conclusively break with the modernist management of technology. The core of modernist management lies in the separation of technology and its social effects. The separation came into being in the early modern period. Resistance by the Luddites can be interpreted as resistance to that separation. They demanded that those who introduced new technology would anticipate the social effects. They called the promoters criminals because of their refusal to take responsiblity for the effects. To the Luddites and their sympathisers technology and its effects were not separate realms. This was the case for the Rotterdam dockworkers. One could only compensate for the effects. No alternative development was possible. Technological development was seen as unavoidable. Environmental groups and other protesters against the prospective expansion of Schiphol were more ambivalent. They attempted to formulate alternatives and did not define the contemporary plans for Schiphol's expansion as unavoidable. At the same time there was little resonance as well as space to work up alternative products for Schiphol to deliver.

The lack of what I call negotiating space between the actors involved in the design process and spokespersons for actors who are directly affected by the technology is a feature of the modernisation process as it has manifested itself till now.[13] In the modern regime of technology management, two tracks are apparent: promotion and regulation. On the one hand there has emerged separate sites – called laboratories – where designers are given plenty of room to tinker with new technologies without having to think about the effects of their introduction. They're not even allowed to think about the effects, for creativity may suffer. After they've been tried and tested, the black boxes are sent off into the world to bring about welfare and progress. Just plug it in; playing with the technology is even considered dangerous, and thus not valued. On the other hand there has emerged a regulatory arena to mitigate the appearance of negative effects. Regulation does not concern itself with steering the scientific and technical developments, but rather with setting limits to their application.

Since the 1970s more and more problems and limitations have become associated with this dual-track approach. Particularly in the decades after the Second World War, people were promised that science and technology would solve their problems. But their salutary promise did not pay out. More and more problems cropped up and so-called negative side-effects of existing technologies were not be solved through ex post regulation. They only worsened. Environmental problems are good examples. In the past twenty years we have witnessed an explosion of new governmental regulations as well as a great increase in knowledge of environmental problems and solutions. Environmental advisory agencies have flourished. Yet the environmental problems appear only to have worsened. Chimney filters and catalytic converters appear unsatisfactory. It has become clear that environmental problems must be addressed through a drastic reduction of energy and resource use. Another form of production and consumption is required; sustainable development is the term used. An alternative form of production and consumption implies not only making environmentally-friendly technologies, but also an alternative form of making technology – according to CTA, as I will argue.

Because science and technology have not delivered their promises, the responsible institutions increasingly have lost their credibility. Experts are no longer automatically taken for their word. Some new technologies, as biotechnology, are being met with new, robust forms of resistance. Nuclear energy already has been effectively frustrated. Paradoxically, science and technology are being called upon to solve the problems.[14] A quick implementation of new technology is now also a political priority. Convincing people to accept new technology is high on the agenda of a government wishing to stimulate technology. Beck (1992) points out that discussions about risk production and risk distribution have come to dominate the political agenda and have become just as relevant as discussions about economic growth and income distribution.

In order to overcome the problem of acceptance, firms are increasingly anticipating the prospective social effects, especially when developing "sensitive" technologies. They enter into discussions with social groups at an early stage. Social groups also sometimes seek contact with the firms, and some governments and agencies attempt to support this kind of cooperation. Researchers and policy-makers attempting to develop the CTA perspective are following this trend and are trying to shape the process. The development of the perspective is to be understood as an attempt to articulate what is going on and which steps could be taken to improve the integration of technology and society. The diagnosis is that such an integration cannot be achieved by undertaking research into effects, for example by doing technology assessment. Rather, the character of the design process is in need of change. It must be broadened to include social aspects and actors. Ultimately such a broadening could lead to a change in the current pattern of technology management (the dual-track approach). New institutions should emerge which will become platforms for the constructive integration of technology and society. It is constructive not in the sense of conflict avoidance, but in the sense that all of the affected are in a position to take responsibility for the construction of technology and its effects. In the existing dual-track regime, no one really takes responsibility for the effects. To technology developers society must see to the effects; they do not follow from the technology. Societal The actors do not feel responsible for the effects; they subsequently call for protection from the government. By institutionalizing CTA practices proponents and opponents both will become responsible for giving meaning to technology and its effects. This will deliver a new politics (or following Beck (1992) subpolitics) of technology.

The Features of CTA

The view that design processes must be broadened is not based on the presumption that social effects play no role in the design process. On the contrary, they are present in the form of (sometimes implicit) assumptions about the world in which the product will function. In technology studies there is the relevant notion of scripts (Akrich 1992, and 1995) wherein the roles of artifacts, actors and the effects are pre-set. The effect of broadening (and thus of the application of CTA) is that the designers' scripts[15] are articulated and laid bare as early as possible to the users, governments and other parties who have their own scripts, and who will feel the effects of the technology. From the point of view of CTA, it's important to make room for such an early and more regular confrontation and exchange of all the scripts. Thus CTA processes acquire their three beneficial features: (1) anticipation, (2) reflexivity, and (3) social learning.

The Importance of Anticipation

Whenever users, social groups and citizens take part in the design processes, they are more likely to bring in social aspects at an early stage than are designers. Designers rarely anticipate social effects; they even have a hard enough time anticipating market conditions in a timely fashion. They do not seem to seek the relevant market information, and even when they do they do not seem to be in the position to put it to good use. They react to market signals and social effects only when they occur, which leads to ad hoc problem-solving (Slaughter 1983; Fleck 1994 and Akrich 1995).

When involving users, social actors, citizens etc. in the design processes in order to anticipate effects, it's important not to structure the process too much in advance. The existing method of user research is to ask the users to react to pre-set product ideas. Users are not invited or given any space in which to come up with their own ideas. Car users, for example, are invited at the so-called "clinics" to try out the latest model for a couple of hours and fill in a prestructured questionnaire. They are not put in a position to define the problems themselves and experiment with various modes of mobility. Consequently consumers seem to ask only for more comfort, speed and acceleration capacity. A different procedure was sought during the traffic discussions in Groningen in 1995. Conversations and discussions were organized between citizens. In four so-called working atelier four visions were developed to achieve an accessible city in future. Significantly innovative and feasible traffic systems were developed (NRC-Handelsblad 18 juni 1996). From the perspective of CTA, it should be noted however, that traffic system designers were not present at the discussions. In order to encourage anticipative behavior in users and designers, it's important to organize societal experiments where participants are stimulated to consider possible synergies between design, market conditions and social effects.

Despite the emphasis on anticipation, there is no presumption here that all social effects can be predicted. On the contrary, it must be assumed that technological development is non-linear and unpredictable. During development all kinds of unexpected sideroads and branchings emerge. Path dependencies will appear; certain solutions chosen for local reasons will continue to drive technological development. But this unpredictability also does not mean that anticipation as such is not possible or is senseless. There are methods now under development which attempt to take into account the non-linear capricious character of technological development and build upon the notion of path dependencies (Rip 1995; Kemp et. al 1998).

The given unpredictability of technological development has two implications. First, anticipation must be organized into a regular activity, also during the phase of implementation. That's when new unforeseen effects emerge by way of new interactions and applications. Owing to the importance of anticipating social effects as early as possible Jelsma and Rip in a recent study (1995) advised firms to organize a trajectory to develop scenarios for coping with social effects, alongside product development trajectories. Second, the technology development process should be structured flexibly so that choices can be deferred or altered. If flexibility and alternation are built into the standard design process, the effect is that the "things" themselves take on the form of an experiment. This way they become more open to input from social groups. Herbold (1995) has shown how the design of a disposalsite could be designed to allow for changes to reduce risk in later stages. This gave designers and social groups the time to negotiate a definition of a safe site.

Reflexivity

Broadening the design process results in being able to notice earlier and more clearly that social effects are coupled to specific technical options and that designers' design not only technological but social effects. Scripts can no longer remain hidden. The effects that emerge are dependent not only on the designers' scripts but also often on the outcomes of complex interactions between designers, users, third parties and the context in which these actors operate. CTA activities aim to stimulate actors to take account of the presence of scripts and realize that technological developments and social effects are co-produced. Actors thereby become reflexive. They must integrate things and their effects into their thoughts and actions.[16]

Consensus may be reached, but controversies could very well occur. CTA could bring about controversy for hidden scripts are exposed and placed next to one another. That needn't be such a great problem in societies where controversies are a normal part of the process of technology assessment. Analyses of controversies made by Jelsma (1995) and Wynne (1995), for example, have shown that often attempts are made to suppress reflexivity. Attempts are made to separate technical facts from assumptions about the social reality in which the technologies function. Controversies subsequently take the unproductive course of the dual-track regime, as I argued with the Schiphol case.

Societal Learning Processes

Technological development can be described as a process where new couplings are forged between technical standards, production structures, market conditions and cultural notions. Thus the development of the electric car can be viewed in terms of the development of new batteries and drive chains; consumers change their mobility patterns in favor of a smaller radius of activity; and society comes to appreciate quiet transport. (Slow electric cars are silent.) In the process actors learn to specify technical standards, the market conditions, the sufficiency of the current institutional structures, etc.[17]

In current design processes learning processes are organized in a linear manner. First there is an attempt to optimize the technology of a process or a product. Second the production requirements are better specified, third the market demand is evaluated, and finally the social effects are taken into account, as they affect the firm. Of course no design process is as linear as this. There is an attempt to work with feedback mechanisms. Feedback also arrives unexpectedly, as problems during application force redesign (Jeslma and Rip 1995). But such adjustments do not greatly change the basic linear character of the process. Social learning with respect to the social effects almost always relates to learning about a technical configuration which is more or less known. This has serious consequences. Not only is it difficult to make larger changes (environmental problems, for example, are solved by installing a filter in a chimney or a catalytic converter in the exhaust system) but the learning process also is determined by this basic structure.

Learning could happen on two levels. The first level relates to developing a better ability to specify and define one's own design. Second-order learning means learning about one's own assumptions and scripts, learning that one is creating new couplings and demands (Deuten and Rip 1996, 11 and Grin and Van de Graaf 1996, 77). CTA relates to both forms of learning. It's important to embed technological development in societal learning processes as early as possible so that users, designers and third parties have the opportunity to scrutinize their own presumptions and come to new specifications. In practice design processes become then more symmetrical from the beginning. As much attention is paid to technical as market and societal issues. Design processes become open (so actors are ready to partake) and space is made for experimentation, for trying out various couplings and problem definitions.

CTA Revisited

CTA activities are not directed in the first instance at such substantive goals as the reduction of environmental pollution, the creation of more privacy, etc. Thus for instance the development of wind energy or a security system to guard against bank fraud cannot be automatically labelled CTA. The purpose of CTA is to shape technological development processes in such a way that social aspects are symmetrically considered. When design processes assume the character of CTA fewer undesired and more desired effects will result. Such a claim is based on two arguments. (1) By incorporating anticipation, reflexivity and social learning, technology development becomes more transparent and more compliant to the wishes of various social actors. They will address the social effects that are relevant to them. (2) In a society where CTA processes have become the norm, technology developers and those likely to be affected by the technology will be in the position to negotiate about the technology. An ability to formulate sociotechnical critique and contribute to design will become widespread. Resistance to specific social aspects will not be viewed as technophobia, but as a signal to take up and as an opportunity to optimize the design (or achieve a better fit in society).

The effect of CTA will not be to bring technology under control so that it plays a less dominant role in society. What changes is the form of control and how technology development is played out. CTA concerns changing change; designs would be designed differently. The goal is to anticipate earlier and more frequently, to set up design processes to stimulate reflexivity and learning, and thus to create greater space for experimentation. Where possible technologies should be made opener and more flexible so users can easily have control over them. Technological development also will become more complex. More coordination and new competencies will be required. In some cases technical change processes will slow down. New institutions will emerge to encourage negotiation between developers, users and third parties. Should design processes acquire the character of CTA, technologists will not suddenly see their work disappear or have it constantly evaluated by all sorts of commissions. Most all of the incremental design changes will not require negotiation. In the program of requirements allowance routinely will have been made for social aspects (including flexibility). However, the variety of technological designs probably will increase, for more groups will be addressed in their capacity as knowledge producer and technology developer (Verheul and Vergragt 1995).

The three quality criteria for CTA processes make apparent that broadening the design process is not an end in itself and that "broader" doesn't necessarily mean "better". Broader is better only in those design processes where space has been created for anticipation, reflexivity and learning. That provides some guarantee that processes should result in better technology, which is to say technology with more positive and fewer negative effects. With the aid of the three features also existing CTA activities can be subjected to a litmus test. They can be evaluated and suggestions can be given for improvement. But what is CTA in practice?

CTA practice can be divided into activities instituted by (mainly Dutch and Danish) TA organizations and by companies, governments and social groups. Most all of these have not been undertaken from a CTA perspective. However, since they explicitly attempt to broaden the design process to account for social aspects they can be considered (and evaluated) along the lines of CTA.

Activities by TA Institutions

In the Netherlands the Rathenau Institute (formerly NOTA) has played an important role in stimulating the CTA perspective. They have supported studies and organized workshops to further develop the CTA idea. In the publication Koersbepaling in 1992 NOTA's goal was spelled out as follows:

"NOTA has as its aim to support the political evaluation of science and technology development and to contribute to decision-making where varied and sometimes opposed interests must be weighed and balanced". (Rathenau 1995)

In their 1994 program the TA process was described as an intervention in an evolving process of social evaluation. The intervention takes the form of research, and the organization of discussion and interaction. The goal is not so much to suggest ready-made solutions, but to clarify questions raised by technology development and to strengthen interaction between various groups who are somehow involved in the development (Rathenau 1995 and Van Eijndhoven 1995). The purpose of "research and discussion" continues the view developed in the first TA governmental report Integratie van Wetenschap en Technologie in de Samenleving (IWTS) in 1984, based in particular on the work of Smits and Leyten (1991).

The work of the Rathenau Institute occurs along these lines. In its various projects research is conducted which places side by side various standpoints; workshops are organized with the participation of various groups who check and further articulate the standpoints. In addition, Rathenau has organized a number of large public debates where in the course of three days a well-prepared panel of 15 laypeople pose questions to experts whom they themselves have invited. The conclusions of the debate are laid down in a closing statement.

The debates organized by the Rathenau Institute follow the consensus conference model, which has been used in Denmark for a number of years as an advisory tool for Parliament. The organization of debate and discussion has a long tradition in Denmark, not only in terms of consensus conferences but also dialogue and scenario workshops. In addition social experiments with new technologies are carried out. The social aspects of new technology can be evaluated in practice (Cronberg n.d.; Remmen 1995, Andersen 1995 and Hetland 1994).

Besides Rathenau, in the Netherlands the SWOKA Institute (Institute for Strategic Consumer Research) takes the approach of research and discussion. Its activities carry the CTA label. They've developed a practical method to practice CTA from a consumer point of view: the future images of consumers (or the TvC procedure). The goal of the procedure is to attune technology development to consumer aspects. The attunement is achieved through the organization of rounds of research, confrontation and deliberation. The procedure is a kind of testbed laboratory. The works involves the articulation of technical standards, market demand and societal conditions (as regulation).

Normative evaluations of the various aspects make up a part of that articulative work. The model has been applied for example in the case of "novel protein foods" by the interdepartmental program, Sustainable Technology Development (DTO), supported by the Ministries of VROM, LNV and Economic Affairs as well as by Gist Brocades and Unilever. The central question was whether it was possible for consumers and producers to create appealing protein foods which displace meat and put less pressure on the environment. Representatives of universities, marketing organizations, supermarkets, firms, consumers, environmental organizations and ministries have met three times. At the third meeting a closing statement was prepared that indicated the standards novel protein foods would have to meet, the market prospects and the further steps the representatives could take (Fonk 1994, SCOPE 1996, no. 2, 6-8).

The crucial question for CTA is the contribution of these communication experiments to the broadening of the design of technology development. Do these experiments couple debate and design? Or is it the case that debate and design only lead parallel lives? Another question concerns the type of coupling. Do processes of technology development thereby become more reflexive? Are they organized so as to encourage anticipation and social learning?

The Dutch and Danish initiatives, in the first instance, are aimed at furthering social debate about science and technology between various social groups in order to report to Parliament. In certain projects recommendations are made by various social groups. Many debates have as their aim the clarification of the political debate and the available options in that context. Bringing forth concrete design solutions is not the most important outcome. Consequently it doesn't seem relevant to pose questions in regards to their effects on the design process. It is appropriate to label these efforts TA and not CTA. Two caveats are in order. In the first place certain Danish initiatives (namely, the dialogue workshops and the social experiments) and SWOKA's work place the accent less on debating than on formulating design specifications for new technological trajectories and learning to cope with new technologies (albeit without new designs). Second, the absence of a direct coupling between debate and design does not necessarily imply that debate has no influence on design. Van Eijndhoven (1995), director of Rathenau, writes:

> "The political process supported by technology assessment has its counterpart in other social processes, as the development of options for firms. Shifts in political opinions go together with shifts in social opinions and in the manner in which products and services are generated and brought to the market".

Such a connection has never been researched, but one could assume it is the case. Rathenau's activities contribute to the articulation of political and social acceptibility (and in some cases to that of market demand) which could influence technology development. Influence is a diffuse process. It emerges through the the participation in Rathenau activities by representatives of the design constituency as they pose questions and arouse expectations. In the biotechnology discussions, for example, firms are present and may influence the new governmental regulations. Because of the debates, reports, workshops, etc. designers are in a better position to anticipate the desires and demands of larger publics as well as social groups. The effects of things are clarified, alternatives are formulated. Through interaction diverse representatives learn to scrutinize the existing desires and demands and eventually define new ones. Thus interpreted the debates are the negotiating space which helps to bridge the gap between the twin tracks of promotion and regulation. It is apparent that the development of this negotiating space is in a precarious phase. It is still organized on an ad hoc basis and does not enjoy much status. But these activities may grow to become normal practice. New networks could emerge to further the processes of articulation. For these reasons the Rathenau and certain similar Danish activities were held up as exemplary by Rip, Misa and Schot (1995).

Design in Practice

The activities of TA institutions are organized by third parties – organizations which are outside of the design process. Another important question concerns whether the actors closer to the design process undertake CTA activities and how these may be evaluated.

In my discussion of the features of CTA I indicated that design processes are largely organized in a linear way and are not of the CTA variety. Moreover, such a linear structure reflects the manner in which technology development is embedded in society. There are separate promotional and regulatory tracks. Such a general conclusion does not imply that there are no initiatives underway to broaden the design process. There are many. They are undertaken by certain firms, social groups and government agencies. They take action because they are confronted with dilemma. Promoters of technology development are confronted with the problem of acceptance. The risks associated with investment are increasing. Thus certain technologies can only be further developed with great difficulty. At the same time technology is still viewed as a problem-solver and a source of competitive strength for countries, industrial sectors and companies. The discussions surrounding the many applications of biotechnology illustrate the dilemma well. There are two types of promotors who undertake broadening initiatives: companies and governmental agencies responsible for technology policy. I begin with the latter.

In Dutch technology policy (of the Ministry of Economic Affairs) the societal entrenchment technology is part of the third phase of technology policy.18 It is thought that through entrenchment policies public support for new technology would grow. Another important motivation concerns the inadequate balance between supply and demand and the consequent waste of resources. Many technological options go unused. By improving that balance the diffusion of new technologies could be smoother. Moveover a number of social problems could be solved. If society better articulates its demands, then the supply could be attuned to the demands, at least according to the reasoning of Dutch technology policy makers. The policy comprises the organization of information activities (to increase public support), the organization of strategic conferences where users, government and providers attempt to identify technological options and related governmental policies to solve social problems, and the financing of experiments with new technologies where users play a large role. An example is Amsterdam's 'Digital City' where users and government communicate with each other. One of the goals is to gain experience so new products and technologies can be developed.

As a part of their product introduction strategies some firms have been pioneers by taking up contact early on with environmental organizations and/ or other social groups, or in some case with members of the local community or other laypeople who feel affected by the new developments.[19] These firms do not wish to wait for the governmental regulations (which can produce legitimation), but prefer to make their own assessments as to which aspects of the products are likely (or less likely) to cause concern. In some cases the activities are not undertaken by individual firms but by groups of firms, sometimes in collaboration with a number of social groups. One example is European Partners for the Environment (EPE), a cooperative group consisting of large firms such as Dow Chemical and Proctor & Gamble, environmental groups, unions, cities and international organizations. In such places as the so-called sustainability laboratories they discuss solutions to mobility problems and environmental problems in agriculture and in tourism. Another example is the Dutch High Definition Television platform, where consumers, professional users, government and providers guide the introduction of high definition TV in the Netherlands.

From the perspective of CTA a number of evaluative remarks may be made about these kinds of activities. In regards to activities by individual firms, often times the social groups are called in too late, and are able only to comment on a number of specific pre-set items. Social learning, reflexivity and anticipation have a limited chance to develop. This obtains for the HDTV platform and most of the activities organized by the Ministry of Economic Affairs. The purpose is to gain greater public support for a specific technology so the introduction of the product becomes smoother. There is little opportunity for redesign. This is less so for the activities of networks as EPE. In those networks there is an attempt to think through existing product bases anew. The formulation of new forms of mobility is on the agenda, as are new technology (electric cars, traffic relief) and new forms of usage (common car ownership and pooling). In these networks responsibility for the development of new technological trajectories as well as new problem and solution definitions is shared.

The evaluation is rough, and will require further development and depth. However, it illustrates the point I wish to make. Initiatives by government, firms and social groups may be viewed as CTA activities. Though sometimes unintended, they do lead to the broadening of design processes (albeit in rather diffuse manners, as I discussed with reference to TA institutions) and they sometimes meet the criteria of CTA processes, that is, anticipation, social learning and reflexivity.

The Future of CTA Research

CTA intiatives will become more widespread. The problems associated with technology are pressing, while at the same time technology development remains an important means to solve problems, earn money and shape society. The question remains whether the current CTA practices will take on the features of proper CTA. Current CTA practices comprise a long series of diffuse efforts to broaden design and are undertaken by a broad spectrum of actors. They still are conducted in an ad hoc manner and are aimed at opening discussion and dialogue. It's not quite clear whether they are coupled to the design process. Actors do not seem prepared to give up their own identity. Social groups and firms do not wish to commit themselves to the outcomes of the interaction; they wish to retain their identities as promoter and regulator. The dual tracks appear firmly rooted. This also was true in the Schiphol affair.

What could guide CTA? First, there should be further evaluation of CTA practices, where the outcomes are discussed with the actors. The evaluations could be directed towards identifying the factors leading to success and failure, and ways to incorporate the CTA features described above. Second, research must unravel the mechanisms by which technology development is influenced. On the basis of research carried out thus far, three kinds of couplings may be defined: (1) the articulation of expectations and needs, whereby new technological options may be identified (or, in the extreme, be forced, as for example when government has legislation passed defining certain future functions of technology); (2) the formation of networks between social groups, users, governments and technology providers; (3) niche formation the creation of protective spaces where all those involved can experiment with new technology (Schot and Rip 1998). CTA researchers needn't only evaluate; from the evaluations they should develop new instruments (as SWOKA's method; see also Weber et. al forthcoming), so actors are in a position to actually implement reflexive processes. Finally, it is important that CTA activities are seen from a long-term perspective. That is one of the intended contributions of this article. That way we gain a better understanding of the working of technology, and embed criteria for CTA in an historical analysis of modernity.

Epilogue

The point of CTA is change. That is why CTA is directed towards intervention. It is not concerned with single point intervention, however, as how actor A can impose his/her will on actor B. CTA processes result in a new politics of technology. That is politics aiming at the development of new broad and heterogeneous networks in which actors develop processes which feature anticipation, reflexivity. In this article I also wanted to point out that CTA extends beyond changing and/or improving the design processes surrounding individual technologies. The point of CTA is ultimately to change the way design is done in our modern society. Change does not imply that the design activity itself needs to be put under discussion. The CTA perspective accepts modern society. That is to say a society where there is room to innovate and create stable artefacts and networks. Only the design process is the object of change. To make that change inspiration can be sought from (portions of) Romantic thought, whereby technology and society are not pulled apart and whereby individual autonomy and the relevance of different truths held by different groups is accepted. The design process must make way for open confrontation and the exchange of truths.[20] Only then will reflexivity, anticipation and social learning be well-served.

Is CTA also aimed at democratising technological development? On the one hand it's quite easy to say that CTA should result in more democracy, because the spread of CTA will lead to more involvement in technology development by citizens and social groups. But that answer is too easily sought. The crucial question is when participation or involvement can be held to be democratic. In CTA participation is not sought in principle to achieve representation or to effect a balance of interests (as in the pluralist variation of democratic theory). Moreover, interests are not viewed as static. Defining interest is exactly one of the efforts of technology development. In the participatory variety of democratic theory, much emphasis is placed on the effect participation has on the participants involved. According to Laird (1993, 345):

> "Truly democratic participation changes the outlooks and attitudes of participants. It makes people more aware of linkages between public and private interest, helps them to develop a sense of justice, and is a critical part of the process of developing a sense of community."

Emphasis on the education of the individual participants is not central to CTA. The purpose is to shape design processes. To shape anticipation, reflexivity and learning. Whenever those features are adequately present, a further broadening of participation is no longer necessary. CTA is concerned first and foremost with the development of better technology, which is to say technology embedded better in society. Improving democracy in a traditional sense (connected to decision-making in the political system) is not the priority of CTA, but may very well be an unintended effect. Formulated in another way: CTA is seeking to establish a new form of democracy wich fits a society full of hybrids and cyborgs.

Notes

1. This article is a revised and translated version of the an article published in Dutch in Kennis en Methode, XX (1996) 3, 265-293. It is also published in A. Jamison, PESTO papers II, Aalborg University 1998. This edited volume results from the EU funded project Public Engagement and Science and Technology Policy Options (PESTO). I would like to thank Richard Rogers for his translation and suggestions and Andrew Jamison, Annemiek Nelis, Ibo van de Poel, Arie Rip, Philip Vergragt and the editors of Kennis & Methode for their comments.
2. Mumford (1963; first edition 1934) 287.
3. Romanticism is a complex movement, the focal point of which lies not in critiques of technology. Romantic authors as Carlyle and Southey in England and Adam Mueller in Germany did, however, formulate a radical critique of industrial society. They were against the fact that industry and technology were not subject to official norms. To them new technology embodied undesirable developments. Later Romantic critique centered on the ugliness and cruelty of industrialism. The prospect of a more humane alternative was lost. But the critique has remained, and it has become a significant counterforce in our culture. (See Sieferle 1984) Yet it lacked a realistic strategy to effect change. Recently there has been some movement in that direction (owing in part to the rise of technologies with many

'effects' and the recognition by promoters that public participation was needed, in order for the public to accept the new developments. Another reason relates to the insight that technology is engineered by people.) A similar conclusion is drawn by Feenberg, (1995).
4. In this article 'design process' and 'technology development' overlap.
5. Much of the CTA literature discusses the expansion of actors or aspects (as in the first CTA report, Daey Ouwens 1987; see also Fonk 1994). Such formulations lack the required precision. The point is that social aspects are introduced in the design process in such a manner that the design processes themselves acquire the quality of constructive technology assessment, as is discussed in this piece. Furthermore, one cannot expect that the introduction of social aspects without the introduction of social actors will lead to an expansion of the design process.
6. See Van Berkel and Stemerding (1998) for an example.
7. For this notion of politics of technology, see Bijker (1995). In this article, Bijkers puts CTA wrongly in a box of policy studies.
8. Overviews of experiences and studies are found in Daey Owens et al., (1987); Rip and Van den Belt (1988); Schwarz and Thompson (1990), Schot, (1991); Schot (1992); Cronberg (n.d.), Remmen (1991), Fonk (1994); Rip, Misa and Schot (1995), Jelsma and Rip (1995) and Schot and Rip (1998). Most of the publications are attempts to formulate and develop a CTA perspective, combined with an analysis of empirical material. Rip, Misa and Schot's book resulted from an international worshop at the University of Twente, supported by the Ministry of Economic Affairs and NOTA (now the Rathenau Institute). Most CTA work till now may be viewed as research, mainly in the form of historical case studies. There is also the practice of expanding the design process, which is not based explicitly on the CTA perspective. An important next step is to evaluate the on-going design projects from a CTA perspective.
9. We have borrowed on some occasions the notion of management, see for example Rip, Misa and Schot (1995). They refer to management not in terms of strategies to achieve a desired result. Instead, management refers to pattern of management or the ways in which a particular society integrate technology.
10. The workers later were given the name 'Luddites' after their legendary leader Ned Ludd, who signed messages in the name of the workers. He probably never existed.
11. Luddism still ocassionally surfaces as a term of abuse, and even as a sobriquet. An example of the latter is that last year the second Luddite Congress was held. (See Financial Times, 15 April 1996, with thanks to Dany Jacobs.) See also Sale (1995). Others use the name Luddite to indicate that technology and its effects should be evaluated in an integrated manner. The subtitle of my dissertation (Schot 1991) thus calls CTA present-day Luddism.
12. Here 'constructive' does not mean the absence of conflict, but rather interaction whereby actors can contribute to the design process.

13. Staudenmaier (1989) has made a similar analysis and he found that in American history negotiation has had to give way to standarisation. Consumers, workers, and citizens have been given less space to adapt (or negotiate with) artifacts. They were forced to organise their lives with standardised products.
14. A 1987 survey, carried out by the Social and Cultural Planning Agency, revealed that 55% of the Dutch surveyed agreed with the position that 'many of the current problems will be solved by new technological discoveries', while at the same time 69% agreed with the position that 'technical experts do not adequately take into account people's actual needs.'
15. I use a broad definition of designers, namely, technicians, managers and workers who are directly involved in the design process. In this respect Staudenmaier 1989 has termed it a 'design constituency'.
16. In Latour's (1994) and Beck's (1992) analyses emphasis is placed on the need for reflexivity. Latour talks of 'making the existence of hybrids explicit' and 'making apparent underground network translations' (chapter 5); Beck speaks of 'self-confrontation with the effects of the risk society' and autonomized modernisation processes which are blind and deaf to their own effects and threats' (p. 6). Keep in mind that here the issue concerns effects on the level of the process and that design processes must encourage reflexive behavior, which is not to say that all actors always must be reflexive. At the same time it's inconceivable that processes can be reflexive and actors not.
17. For a description of electric transport in terms of societal learning, see Schot, Slob and Hoogma (1995). They divide the learning process into six parts: the articulation of demand, of critique, of the social and political preconditions, of the production and maintenance network and of the effects. See also Van Berkel and Stemerding (1998).
18. See the governmental discussion report Technologie en Samenleving, published in 1991 by the Ministry of Economic Affairs, and also later reports as Concurren met Kennis (1993) and Kennis in Beweging (1995) where the 'policy of attunement' is taken up. The expansion of technology policy has become a broader trend in Europe; see OECD (1988). It also is in evidence in the European Commission's programs as MONITOR and VALUE II, which were set up in part with the idea that social aspects are important and relevant for technology policy. In the U.S. calls also are being made. See Branscomb (1993), and for example the work of the Carnegie Commission (1992).
19. This trend is underdocumented. My thoughts are based in part on experience gained in the Greening of Industry Network, founded in 1991. In this network industry, government, social groups and researchers discuss environmental problems. The involvement of stakeholders in industrial policy and R&D is an important theme. See Simmons and Wynne (1993); Chess and Lynn (1996); and Schot (1995).

[20.] In this regard CTA may be seen as an attempt to combine the elements of the Englightenment with certain of those from Romanticism. For an attempt to think through such a combination in the history of ideas, see Berlin (1992).

Literature

Akrich, M. (1992) 'The description of technical objects', in: W.E. Bijker en J. Law, *Shaping technology/building society: studies in sociotechnical change*, MIT Press, p. 205-224.

Akrich, M. (1995) 'User reprensentations; practices, methods and sociology', in: A. Rip, Thomas J. Misa en Johan Schot, *Managing technology in society. The approach of constructive technology assessment*, Pinter Publishers, p. 167-184.

Andersen, I-E. (red.) (1995) *Feasibility study on new awareness initiatives*, The Danish Board of Technology.

Beck, U. (1992) *Risk society. Towards a new modernity*, Sage.

Beck, U. (1994) 'The reinvention of politics; towards a theory of reflexive modernisation', in: U. Beck, A. Giddens en S. Lasch, *Reflexive modernisation. Politics, tradition and aesthetics in the modern social order*, Polity Press.

Canini, G., A. Candaele, J.C. Burgelman (1995) 'Publiekparticipatie bij de besluitvorming over wetenschap en technologie. Een nieuwe uitdaging voor publieksvoorlichting', in: W. Van der Meer en L. Van Langehove (red.), *Management van Wetenschap*, Uitgeverij Lemma, p. 161-181.

Chess, C., en F. Lynn (1996) 'Industry relationships with communities; business as usual', in: Groenewegen, P., K. Fischer, E.G. Jenkins en J. Schot, *The Greening of Industry Resource Guide and Bibliography, Island Press*, p. 87-110.

Cronberg, T. *Technology Assessment in the Danisch Socio-Political Context*, Report no 9 published by the Unit of Technology Assessment, Technical University of Denmark, n.d.

Daey Owens, C., P. van Hoogstraten, J. Jelsma, F. Prakke en A. Rip. (1987) *Constructief Technologisch Aspectenonderzoek. Een verkenning*, Den Haag: NOTA voorstudie no. 4,

Deuten, J.J. en A. Rip. (1996) R&D Management for Public Acceptance: An Important Aspect of Total Quality, paper presented at R&D Management Conference, 'Quality and R&D', Univerity Twente Enschede, 6-8 March.

Elzen, B., R. Hoogma en J. Schot. 1996. *Technologiebeleid voor verkeer en vervoer, Ministerie van Verkeer en Waterstaat*.

Eijndhoven, J.C.M. van. (1995) *De ondragelijke lichtheid van het debat*, inaugurale rede, Rathenau Instituut.
Eyerman R. and A. Jamison. 1991. *Social Movements. A Cognitive Approach*. Polity Press.
Feenberg, A. (1995)*Alternative Modernity. The Technical Turn in Philosophy and Social Theory*, University of California Press.
Fleck, J. (1994) 'Learning by trying: the implementation of configurational technology', Research Policy, 23, p. 637652.
Fonk, G. 1994. *Een constructieve rol van de consument in technologie-ontwikkeling. Constructief Technologisch Aspectenonderzoek vanuit consumentenoptiek*. Academisch Proefschrift Universiteit Twente; uitgegeven als SWOKA onderzoeksrapport nr. 166.
Grin, J. en H. van der Graaf. (1996) 'Technology Assessment as Learning', *Science, Technology and Human Values*, 21, p. 7299.
Herbold, R. (1995) 'Technologies as social experiments. The construction and implementation of a high-tech waste disposal site', in: A. Rip, Thomas J. Misa en J. Schot, *Managing technology in society. The approach of constructive technology assessment*, Pinter Publishers, p. 185-198.
Hetland, P. (1994) *Exploring hybrid communities. Telecommunications on trial*, Ph.D. thesis, Hedmark College.
Hobsbawn, E.J. (1952) 'The Machine Breakers', *Past and Present*, p. 57-70.
Jelsma, J. (1995) 'Learning about learning in the development of biotechnology', in: A. Rip, Th. J. Misa en J. Schot, *Managing technology in society*, Pinter Publishers, p. 141-166.
Jelsma, J., en A. Rip. (1995) *Biotechnologie in bedrijf. Een bijdrage van constructief technology assessment aan biotechnologisch innoveren*, Rathenau Instituut.
Kemp, J., Johan Schot and Remco Hoogma (1998) 'Regime Shifts through Processes of Niche Formation. The Approach of Strategic Niche Management' *Technology Analysis & Strategic Management*.
Laird, F.N. (1993) 'Participatory analysis, democracy, and technological decision-making', *Science, Technology and Human Values*, 18, 3, 341-361.
Latour, B. (1994) *Wij zijn nooit modern geweest. Pleidooi voor een symmetrische antropologie*. Van Gennep.
Lente, D. van. (1988) *Techniek en ideologie. Opvattingen over de maatschappelijke betekenis van technische vernieuwing, 1850-1920*, Wolters-Noordhoff.
Lente, D. van. forthcoming. 'Dutch Conflicts: The Intellectual and Practical Appropriation of a Foreign Technology' in: M. Hard and A. Jamison.

The Machine in the Mind: Technology Discourses, 1900-1939. MIT Press.

Leys, M. (1995) 'Technology Assessment: wetenschappelijke kennisoverdracht als probleem', in W. Van der Meer en L. Van Langehove (red.), *Management van Wetenschap*, Uitgeverij Lemma, Utrecht, 137-159.

Mumford, L. (1963) *Technics and Civilization*, Harcourt Jovanovich Publishers, (First edition 1934).

OECD (1988) *New technologies in the 1990s: A socio-economic strategy*.

Rathenau Instituut (1996) *1986-1996: van technologisch aspectenonderzoek tot maatschappelijk debat, 10 jaar NOTARathenau Instituut*.

Remmen, A. (1991) 'Constructive Technology Assessment', in: Tarja Cronberg et al. *Danish Experiment - Social Constructions of Technology*, New Social Science Monographs,185-200.

Rip, A. en H. van Belt (1988) *Constructive technology Assessment: possibilities and contraints*, Intern rapport Universiteit Twente.

Rip, A. (1995) 'Introduction of new technology: making use of recent insights from sociology and economics of technology', *Technology Analysis & Strategic Management*, 17. 4, p. 417-431.

Rip, A. Th. Misa and J. Schot (1995) *Managing technology in society. The approach of constructive technology assessment*, Pinter Publisher.

Sale, K. (1995) *Rebels Against the Future*. AddisonWesley Publishing Company.

Schot, J. (1991) *Maatschappelijke sturing van technische ontwikkeling. Constructief technology assessment als hedendaags Luddisme*, Proefschrift Universiteit Twente.

Schot, J. (1992) 'Constructive technology assessment and technology dynamics: the case of clean technologies', *Science, Technology and Human Values*, 17. 1, p. 36-56.

Schot, J., R. Hoogma en E. Elzen (1994) 'Strategies for shifting technological systems. The case of the automobile system', *Futures,* 26. 10, p. 1060-1076.

Schot, J. (1995) 'Schiphol, kijkdoos van de moderniteit', in: *Dossier Schiphol. Over economie en ecologie in Nederland distributieland*, rapport gepubliceerd door Vereniging Milieudefensie.

Schot, J., A. Slob en R. Hoogma (1995) *De implementatie van duurzame technologie als een strategisch niche management probleem*. Rapport uitgegeven door het Interdepartementale Onderzoeksprogramma Duurzame Technologische Ontwikkeling.

Schot, J. (1995) 'Milieu en strategie', in: F.L.M. Braakhuis, M. Gijtenbeek, W.A. Hafkamp, *Milieumanagement: van kosten naar baten*, Samson H.D. Tjeenk Willink, Alphen aan den Rijn, p. 25-47.

Schot, J., en A. Rip (1998) 'The Past and Future of Constructive Technology Assessment', *Technological Forecasting and Social Change*. 54, 251-268.

Schwarz, M., en M. Thompson (1990) *Divided we stand. Redefining politics, technology and social choice*, Harvester Wheatsheaf.

Slaughter, S. (1983) 'Innovation and learning during implementation: a comparison of user and manufacturer innovations', *Research Policy*, 22 p. 81-95.

Sociaal en Cultureel Planbureau (1988) *Publiek en techniek*, Cahier nr. 57.

Sieferle, R.P. (1984) *Fortschrittsfeinde? Opposition gegen Technik und Industrie von der Romantik bis zur Gegenwart*, Beck Verlag.

Simmons, P., and B. Wynne (1993) 'Responsible Care: trust, credibility, and environmental management', in: K. Fischer, en J. Schot, *Environmental strategies for industry. International perspectives on research needs and policy implications*, Island Press, p. 201-226.

Smits, R. en J. Leyten (1991) *Technology Assessment. Waakhond of speurhond*, Kerckebosch, Zeist.

Staudemaier, J.M. (1989) 'The Politics of Successful Technologies', in: R.C.

Thompson, E. P. (1963) *The Making of the English Working Class*, Penguin.

Verheul, H. en P. Vergragt (1995) 'Social Experiments in the development of environmental technology; a bottom-up perspective', *Technology Analysis & Strategic Management*, 7 p. 315-326.

Weber, K.M., R. Hoogma, B. Lane and J. Schot. forthcoming. *Expanding Technological Niches. How to manage the uptake/development of sustainable transport systems*.

Wynne, B. (1995) 'Technology assessment and reflexive social learning: observations from the risk field', in: A. Rip, Th. J. Misa en J. Schot, *Managing technology in society*, Pinter Publishers, p. 19-36.

Contributors

Denmark (Aalborg University)
Andrew Jamison, Arne Remmen

Italy (Bocconi University)
Mario Diani, Marco Guiliani

Lithuania (Kaunus University of Technology)
Leonardas Rinkevicius

Netherlands (University of Twente)
Jose Andringa, Robbin te Velde, Johan Schot

Norway (Norwegian University of Science and Technology, Trondheim)
Lise Kvande, Per Østby

Sweden (University of Lund)
Kees Dekker, Magnus Ring, Arni Sverrisson

United Kingdom (Lancaster University)
Patrick van Zwanenberg, Brian Wynne